I0048809

INNOVATION PORTFOLIO

Five Verbs Shape Your Team's Legacy

ROBERT F. SNYDER

PRAISE FOR *INNOVATION PORTFOLIO*

Innovation Portfolio is refreshing and fantastically provocative. It clarifies behaviors that hurt culture and clarifies habits that shape a disciplined and empathetic culture.

– Line Karkov, business analyst and "Yes And" evangelist

Innovation Portfolio is a recipe to improve your team engagement, morale, and project success rate. The tools provide a new, clean baseline for team expectations. Documentation is not just a formality; it is a necessity. Without a good balance of documentation, you risk wasting time, money and resources. I'm excited to see how Five Verbs will give my team clarity, confidence, and courage to reshape our culture.

– Siu Chu, business systems analyst and IIBA chapter officer

In *Five Verbs*, Robert offers an opportunity to re-think not just what we are doing, but why we are doing it. The simplicity of Five Verbs keeps us focused - ruthlessly focused - on doing only what matters. This clarity minimizes wasteful, churn-inducing, ambiguous labor. Elegance elevates brainpower from the mundane into strategic creativity.

– Ben Kleinman, strategy and change management consultant

Many great books "go deep," and a few "go wide," but *Innovation Portfolio* manages to do both while remaining entertainingly readable, immediately practical, and wonderfully innovative itself, as you'd expect. For innovation organizations averse to documentation, this book will change your mind.

– Andrew Sykes, professor on sales, habits, and trust

Innovation Portfolio is a breath of fresh air. "Hey, it's Agile," has too often been a rationale for reactive decisions and lack of accountability. Snyder's combination of ruthless discipline and tangible empathy is the right formula for change agents wanting to stop the pain of failing programs, to create better business outcomes, and to improve the employee experience.

– Rob Asen, consulting practice executive and culture champion

This book is a revolutionary guide for every team that has ever felt bogged down by meetings and emails. *Innovation Portfolio* brilliantly elucidates the transformative power of discipline, empathy, and attention to detail for teams to overcome the VUCA-ness of today's world. A must-read for those ready to shift from the chaos of 'VUCA' to a world of elegance and efficiency.

– Oge Nwachukwu, project manager, business analyst, change manager, and organizational transformer

Innovation Portfolio overcame my initial skepticism that I was reading about another methodology the world didn't really need. I felt advocacy for the Five Verbs concept the moment I read it. Robert convincingly articulates why appropriate, effective, and durable documentation is more than a necessity. It provides the necessary professional courtesy and kindness for people-centred change. *Five Verbs* is a beautifully simple set of actions to curate documentation assets and promote transparency, alignment, and impact. This book is a must-read for experienced and emerging professionals.

– Jamie Toyne, founder & CEO of Herd Consulting, a specialist business analysis consultancy

INNOVATION PORTFOLIO: FIVE VERBS SHAPE YOUR TEAM'S LEGACY
Copyright © 2024 by Robert F. Snyder, Innovation Elegance, LLC

All rights reserved. This book is for educational and informational purposes only. The author's views are personal and not authoritative. Each reader is responsible for their actions, and the author and publisher are not liable for any outcomes resulting from the use of this material.

No guarantees of success are made, and readers should seek professional advice for specific situations. Any resemblance to real persons or events is coincidental, as the content is fictional or imaginative.

ISBN:
Hardcover: 979-8-9888997-3-0
Paperback: 979-8-9888997-4-7
eBook: 979-8-9888997-5-4

Interior Formatting by Edge of Water Designs, edgeofwater.com
Cover Design by Mariana Coello

Publisher: Innovation Elegance, LLC. Chicago, USA
innovationelegance.com
First edition.

Dedication

My first dedication of this book is to the performing artist in all of us. The serious, competitive world needs more of how we listened while making music, rehearsed in a school play, and mastered the dance floor. That "more" includes the joys of learning, empathy, and being in an ensemble.

My second dedication of this book is for innovation professionals with an appetite for detail. The vulnerability and clarity you bring to teamwork improves relationships, mental health, job satisfaction, and project success rates.

**Other Books by Robert F. Snyder
in the Innovation Elegance Series**

*Innovation Elegance:
Transcending Agile with Ruthlessness and Grace*

*Elegant Leadership:
Distinguishing The Good, Bad, and False*

Acknowledgements

Thank you to the numerous professional associations that fed and inspired this work including:

IIBA (International Institute of Business Analysis)
    BA Café (Brussels chapter)
    Brown Bag (United Kingdom chapter)
ACMP (Association of Change Management Professionals)
    Global Connect Forum
    Midwest Chapter
PMI (Project Management Institute)
    Chicagoland chapter
APMG (Association for Project Management) International
    Level Up Panel Discussions (staff and fellow panelists)
AITP-Chicago (Association of Information Technology Professionals)
IMC (Institute of Management Consultants)
    Chicago chapter
IVY
ECN (Executive Council Network)
KACC (Kellogg Alumni Club of Chicago)
Conscious Capitalism
Chicago Innovation
The Executives Club of Chicago

Contents

Preface

Some people make history. Others watch it on Netflix.

~ Unattributed

Over the past few years, I've had the opportunity to listen to numerous innovation professionals who hold titles such as project manager, change manager, and business analyst. Their complaints have a lot in common and have not changed much over the course of my thirty-year career. As this book took shape, it was easy to conclude that existing methodologies were not solving their problems. And why should they? Agile and Hybrid were formulated for software, not for people, and not for these problems.

What makes innovation teamwork difficult is people, not software, and so a methodology formulated for people (and detailed in this book) is not revolutionary—it's just overdue. After an infinite number of lessons learned exercises and social media posts, those lessons have not been fed back into management books or methodology definitions to change systemic behaviors. Systemic (and systematic) behaviors remain highly tolerant of low discipline and low empathy—in a word, messiness. The innovation world even embraced an acronym for this messiness—VUCA—which stands for volatility, uncertainty, complexity, and ambiguity. Many innovation leaders have surrendered to messiness and to VUCA.

This people-centric methodology doesn't preach perfectionism. It navigates human errors with poise, grace, and resilience. It systematizes—and shapes a culture of—empathy. On the other hand, the methodology is ruthless toward systemic and systematic errors. It proactively minimizes the most common bad habits of innovation teams, that is, it aims to solve the complaints among the aforementioned three professions. The methodology systematizes—and shapes a culture of—discipline.

Formulating a people-centric methodology is anti-climactically straightforward. It finds inspiration from three familiar metaphors (explained in the introduction). But *straightforward* is not the same as *easy*. To be worthwhile and profitable, these metaphors are demanding. They demand that the participants stay attentive to each other. Adopting this methodology rejects low discipline and empathy. It says yes to high discipline and empathy.

At first, discipline and empathy might feel like opposites. Discipline can feel harsh, and empathy can feel loose. But combining them to manage and mentor team behavior leads to a breathtaking employee experience. Discipline and empathy shape a culture of elegance that is not harsh or loose. On the contrary: VUCA, messiness, and chaos are harsh and loose. When teams conquer VUCA and messiness, they collaborate better, they compete better, and they delight their customers. This requires attentiveness, language, and habits embodying elegance.

It can be a delight to observe team elegance from the outside but experiencing it from the inside as a team member is even more exhilarating, especially for the professions of project manager, change manager, and business analyst. Executing this people-centric methodology will result in exhilarating collaboration, repeated project successes, and durable change. The results are a portfolio of improved customer and employee experiences. As innovation professionals navigate their careers among numerous organizations, they will have much to be proud of—positive legacies of customer value and employee experiences. These portfolios and legacies outlive our projects, relationships, and careers. Elegance is a people-centric methodology because the portfolios, legacies, and *people* are worth it.

INTRODUCTION

The cross-pollination of disciplines is fundamental to truly revolutionary advances in our culture.

~ Neil DeGraase Tyson (b. 1958), American astrophysicist and author

I nnovation Elegance—a methodology to govern innovation teamwork—springs from a cross-pollination of three metaphors: a factory, the empathetic arts, and an asset portfolio. Volume One of the Innovation Elegance series defines the methodology using the culture traits of a factory and the arts for comparison. However, Volume One stays shallow about the third aspect of the methodology: the documentation. The various documents built using the Elegance methodology hold their value so well that they qualify as "assets" for the organization, thereby establishing an innovation team's asset portfolio. This Volume Two provides details of the why, what, and how of these assets. This book is instrumental for those who want to get "into the weeds" and use these assets as part of their innovation strategy.

As a foundation for the asset portfolio metaphor, Volume Two leverages a few lessons from Volume One. The first lesson is that a team-centric methodology is more valuable than a software-centric methodology. Software-centric methodologies claim what makes innovation difficult is software. In comparison, the team-centric Elegance methodology claims what makes innovation difficult is people: how we interact, govern each other, compete, and collaborate. A team-centric methodology shapes culture—where culture is defined as "shared behavior"— to make teamwork less difficult. The methodology keeps technology, data, and content in service of people, processes, and context.

The second lesson is that a factory can be a metaphor for how innovation teams work. A factory shapes a culture of movement, rhythm, and reliability—at scale. A factory's motions are efficient, automatic, and low cost. A healthy team is a kind of agreement factory.[1] Teams should see themselves as an *agreement* factory, an *alignment* factory, a *decision* factory, or, for the most tentative, an *expectation-setting* factory.[2] Like a factory, an innovation team needs infrastructure to process intense communication traffic.

The third lesson is about applying the framework of Five Verbs®—*draft,*

---

[1] *Agreement* here has a positive connotation, not to be confused with groupthink or mindless consensus.

[2] Examples include expectations about future state processes, assignments, and schedules.

review, revise, approve, and *distribute.* Five Verbs clarifies what team members do in an agreement factory.[3] Their work can't be random or arbitrary, and the output of their teamwork can't be disposable or wasted. The Five Verbs framework that governs teamwork converts documentation into assets with extraordinary—one might say ruthless—discipline and durability. When a team has a rhythm of generating numerous assets, it qualifies as an *asset factory.* Managing just five verbs—and not managing other motions—is straightforward, and it keeps costs low.

Creating an asset portfolio is painstaking work. It requires unusual discipline and unambiguous, shared behavior governed by Five Verbs. It relies on a simple and scalable process that imitates a factory and produces durable outputs. It demands attention to detail and "getting into the weeds."

---

3   As artificial intelligence (AI) grows in relevance, companies need to consider how AI tools might act as team members. Appendix A comments on the current strengths and limitations of such tools.

The Case for an Asset Factory

We are what we repeatedly do. Excellence, then, is not an act, but a habit.
~ Will Durant (1885–1981), American historian and philosopher

Many companies and innovation teams are effectively meeting or email factories instead of asset factories. Meetings are a comfort zone because they provide a sense of inclusion and instant gratification to the participants. Emails are an addiction because they allow for quiet thinking and control over personal responsiveness. Emails do generate a paper trail, but even in overwhelming volume, they risk leaving people out. At their worst, emails fuel a spirit of CYA (cover your "backside").

When innovation is rare and casual, the high marginal costs of meetings and emails are tolerable. Many typical project teams acknowledge that their problems relate to communication; however, doubling down on meetings and emails only compounds communication traffic jams, employee fatigue, and project failure. As innovation intensifies, project teams must shift away from meeting and email factories (high marginal cost) toward asset factories (low marginal cost).

An asset factory retains the most durable aspects of meeting and email factories and rejects the disposable elements. An asset factory replaces the fleeting gratification of meetings with small wins and a sense of accomplishment for the team. An asset factory swaps the ping-pong typing fest of an email factory with a stable sequence of questions every innovation team must answer. An asset factory enables quiet thinking and replaces the culture of CYA with a spirit of contribution, collaboration, and a campfire of ideas.

The quantity of a team's meetings and emails is evidence that these "factories" contribute to overcommunication, impulsiveness, and interruptions. Teamwork becomes impossible to synchronize in meeting and email factories. In contrast, an asset factory's size and scope are manageable. The factory synchronizes sixty assets through Five Verbs. An asset factory is a rich harmony of teamwork that is visible and straightforward.

Meeting and email factories scatter a company's institutional knowledge across countless minds and email accounts. This sprawl harbors ambiguity, undermines alignment, and causes team members to rehash decisions. But an asset factory turns the screws on ambiguity by promoting alignment in every asset and making decisions transparent. An asset factory reduces several costs such as orienting new employees, reassigning them, pausing and restarting projects, and delegating work to junior employees. In meeting and email factories, these activities are messy and laborious.

An asset portfolio is vital, yet documentation-heavy teams reek of the out-of-fashion methodology called Waterfall. However, dismissing Waterfall merely because it generates too much documentation is simplistic. Waterfall originated in an era when the business world viewed software as what made innovation difficult and when modest project scope was rarely a goal. We now know that people are what makes innovation difficult. We also know to avoid risky "big bang" projects and that current state process documentation helps keep project scope modest.

Even though well intentioned, the Waterfall methodology fails to bring out the best of twenty-first-century project teams. What complicates innovation is people, not technology. Although transparency can be unpopular, it is essential in order to reduce risk and improve collaboration. Companies can avoid Waterfall's shortcomings by adopting language, habits, and structures that are people-centric, not technology-centric, and by being ruthlessly transparent about the current state to facilitate small, low-risk projects.

The popular methodology of Agile advocates "projects" (sprints) that are small and low risk, so on the surface, it appears to fix one of Waterfall's shortcomings. However, valuable documentation—the assets in the asset portfolio—is a common if unintentional victim. What fills this documentation void is meeting gridlock, email overload, and communication traffic jams—all severe problems in innovation teamwork. Neglecting documentation is not a people-centric way of doing business. Doing so complicates teams' collaboration and cripples future teams' learning curves. Meanwhile, proper documentation managed in a simple way as an asset portfolio, makes collaboration and information sharing easier. Structured documentation in an asset portfolio is simple without being simplistic. Therefore, synchronizing the proper documentation at a sustainable pace—as an asset factory does—keeps projects small to address Waterfall's flaws and avoid Agile's high tolerance for communication traffic jams. An asset portfolio improves speed and durability and reduces laboriousness, latency in information sharing, and risk.

Meeting and email factories are slow, sloppy, and expensive; asset factories are fast, elegant, and cheap. The Innovation Elegance methodology will help you run an asset factory and maintain an asset portfolio.

Weeds and Attention to Detail

*The difference between something good and something great
is attention to detail.*

~ Charles Swindoll (b. 1934), American pastor, author, and educator

For some businesspeople, "getting in the weeds" has a negative connotation. The phrase treats detailed work as unsophisticated, unimportant, and undesirable. But many job descriptions ask candidates to demonstrate "attention to detail," so working in the weeds with the details does have value. Job descriptions don't say sloppiness is a disadvantage for candidates, but they effectively say that elegance is an advantage.

For innovation professionals, working in the weeds equates to familiarity and faith in your asset portfolio. Innovators who get into the weeds are exposed to and gain confidence about what durable teamwork looks like. Like any form of literacy, this "innovation literacy" increases learning, influence, and value. Literacy opens new opportunities.

Innovation literacy is empowering because it shows how to get out of Methodology Debt[4]®—the state where the methodology prevents a team from reaching its innovation (and thus financial) potential. Once an organization is out of Methodology Debt, its employees have the freedom to innovate at their organization's frontier. Whereas innovation illiteracy keeps a team in debt, innovation literacy enables a team to get out of debt.

Just as spoken languages have different levels of fluency, innovation has different levels of literacy. Low innovation literacy is the mere acknowledgment that innovation teamwork benefits from specific documentation, that is, the asset portfolio. High literacy is a basic understanding of every asset in the portfolio and the relationships among the assets. This book aims to maximize your literacy of the asset portfolio. And although a book can't provide exhaustive examples of completed assets, this book provides asset content where good examples are hard to find.

A team that executes with discipline and empathy must execute with attention to detail. Detail is conducive to innovation; conversely, aversion to detail undermines innovation. Detail requires structure. Opponents of detail see structure as rigidity that limits creativity and freedom. Proponents of detail consider the absence of detail to be ambiguity and sprawl. Proponents see structure as clarity and boundaries that are movable, not rigid.

Project teams can learn something new at any time. Instead of being at

---

4   Methodology Debt is a word play similar to technology debt. Technology debt cuts corners in code to achieve short-term goals at the expense of long-term goals. Methodology debt cuts corners in a team's more general, formalized habits to achieve short-term goals at the expense of long-term, formal habits.

the mercy of volatility, uncertainty, complexity, and ambiguity (VUCA), the asset portfolio helps you methodically and confidently recognize surprises, diagnose their impact, and propagate new information among all assets.

With approximately sixty assets and five verbs, the asset portfolio provides structure, clarity, and boundaries. It is disciplined and empathetic; rigorous without being rigid. It is simple without being simplistic; and straightforward without being sloppy. Its attention to detail is both ruthless and graceful.

The Real Work

> *The best preparation for work is not thinking about work,*
> *talking about work, or studying for work: it is work.*
>
> ~ William Weld (b. 1945), American attorney, author, and politician

The asset portfolio is not just paperwork. For an innovation team, it is *the* work—the real work that has durable value. Meetings and emails are valuable in teamwork. But *old* emails and *distant memories* of meetings depreciate to the point that they are disposable. Every approved asset in the asset portfolio is valuable weeks and months later. The asset portfolio does not lose value.

The language that describes the work on each asset matters. Word choice must maximize healthy collaboration, accountability, alignment, and a conclusion among every asset's contributors. This collaboration is a mandate—not an option. The word choice that governs the work shapes the collaboration.

Project work is best governed by an asset called a Project Plan (detailed in this book). A terrible Project Plan contains complex and ambiguous language, whereas an elegant Project Plan contains language that is simple and ruthlessly clear. A terrible Project Plan reflects the exotic and expansive vocabulary of a team. It's painful to inspect it closely. An elegant Project Plan contains words that maximize the speed and quality of the teamwork and minimize its waste. The vocabulary is limited, and it's easy on the eyes.

The language for this elegant Project Plan must fit with its other components, such as dependencies, assignments, duration estimates, and percent complete calculations. The language that fits a Project Plan is an action word—a verb. But applying an expansive vocabulary of verbs results in Verb Sprawl®—an ambiguous and overly complex Project Plan. Limiting the vocabulary of verbs generates an elegant and unintimidating Project Plan. The Elegance methodology limits verbs in a Project Plan to five—draft, review, revise, approve, and distribute.

With few exceptions (noted later), Five Verbs applies to every asset. For example, for a process flow asset, the team drafts it, reviews it, revises it, approves it, and distributes it. You can't skip or eliminate one of these verbs. Documentation created in other ways will lack collaboration (it wasn't reviewed and revised), lack alignment (it wasn't approved), or sit on a shelf (it wasn't distributed).

Five Verbs is simple and clear. Using them to organize your agreement factory and build your asset portfolio is straightforward. There is no need to formalize work that requires a sixth verb. Five Verbs is where all the work gets done. Five Verbs *is* the action.

Five Verbs doesn't guarantee work of high quality. It guarantees work of high integrity and durability. Work governed by anything other than Five Verbs might be high quality, but also is inevitably ill-timed and disposable. Divergence from Five Verbs exposes a failure to collaborate and align. It is a reason to pause work. Divergence is a sign of poor leadership, followership, and stewardship of company resources. This sounds harsh, because it is—innovation work requires ruthless discipline. There is simply no reason to diverge from Five Verbs in formal planning. A disciplined, courageous innovation team either executes Five Verbs or pauses the work.

Skeptics might dislike the transparency of Five Verbs. They might claim it feels like micromanaging. But Five Verbs doesn't prohibit other work. Innovation Elegance simply views other work as not worth documenting, collaborating on, approving, or sharing. That other work is disposable and not valuable, especially months later.

Outputs of Five Verbs are not rigid. Teams learn new information during a project. A decision-maker might change their mind. When this happens, the team identifies the most upstream asset affected, reapplies Five Verbs (minus *draft* this time), and propagates the new information through the downstream assets. This might not be easy, but it is ruthlessly straightforward. Five Verbs doesn't impose rigidity or fear. Instead, it encourages transparency, accountability, and alignment.

Disciplined and empathetic innovators should be encouraged and emboldened by the Five Verbs framework. When you and your team are executing Five Verbs, your work is valuable and durable, and it has high integrity. When not executing Five Verbs, you're generating work that has none of those qualities. The clarity in both scenarios offers innovators peace and sanity. Five Verbs governs collaborators' real work.

Q&A

Life punishes the vague wish and rewards the specific ask. After all, conscious thinking is largely asking and answering questions in your own head. If you want confusion and heartache, ask vague questions. If you want uncommon clarity and results, ask uncommonly clear questions.

~ Tim Ferriss (b. 1977), American entrepreneur, author, and lifestyle guru

The structure and language of each asset effectively pose questions for your innovation team to answer. Five Verbs forces your team to collaborate on answers to the questions, requiring them to draft, review, revise, approve, and distribute their answers. Understanding how the asset portfolio is constructed and why you should adopt the entire asset portfolio is key to successfully employing the Elegance methodology.

Innovation literacy equates to intimacy with questions and answers (Q&A), especially knowing the questions, their sequence, and their interdependence. But omitting a question by neglecting an asset doesn't make the need for the information disappear. Procrastinating answering questions or leaving them unanswered just creates ambiguity. Without unified, documented assets, answers hide from your team. Why? Because divergent and contradictory responses reside in countless employees' minds across the organization. Omissions represent willful ambiguity and documentation debt. Omissions keep you in Methodology Debt—away from your innovation frontier and freedom.

When viewing innovation teamwork as a factory that synchronizes Q&A, early visibility and confidence in the questions seems advantageous. But skeptics of visibility exist who disapprove of what they call "prescriptive" methodologies. Skeptics' desire for *responsiveness* is well intended but easily slips into being *reactive*—neglecting work you already know is valuable. Q&A skeptics campaign for freedom, experimentation, and working software, whereas Elegance endorses discipline, empathy, and working teams. Skeptics accept meeting and email factories, but Elegance favors asset factories and asset portfolios. Skeptics prefer to provide answers before or without questions. Elegant teams possess a stable suite of questions they answer for every project and across projects. This habit is a form of muscle memory and automation.

Becoming a Ruthless Asset Factory

Don't worry about getting it right. Just get it started.

~ Marie Forleo (b. 1975), American author, motivational speaker, and life coach

The typical innovation organization has habits and a culture that can benefit from increased discipline and empathy. Thoughtfully and at a pace within your team's comfort zone, adopt one asset at a time. Regardless of your seniority, executing this agreement factory gradually converts your organization into an asset factory. Moving away from a meeting or email factory and becoming an asset factory doesn't have to be dramatic or revolutionary. But the transition should not be casual. The best shift toward an asset factory is deliberate.

If your organization mandates specific documentation, propose small ideas and ask for permission before doing anything differently. But if your culture includes safety, receptivity, and some liberty for new ways of working, the risk is low when introducing assets, and forgiveness is easy. Small changes related to documentation are reversible. If an asset is perceived as counter-productive, it's straightforward to discontinue the asset. In a healthy culture, the worst case is that your colleagues reject an idea; you know you tried, and you know what might succeed another time.

It's easier to adopt assets when they are first embraced by senior employees, but this isn't obligatory. First adopters can be junior employees. Furthermore, over weeks and months, the assets *promote* the self-sufficiency of junior employees. Understanding and faith in an asset factory help inexperienced project managers confidently synchronize teamwork. Planning, assigning, and doing the work becomes straightforward—especially the time-consuming steps of drafting and revising assets. Delegating work to junior employees keeps the marginal cost of innovation low.

No methodology, including the Elegance methodology, is static. But this book's comprehensive explanation of the timeless asset portfolio helps your team minimize "reinventing the wheel," that is, identifying new assets for each project. That said, inevitable improvements to the methodology will grow and reshape your organization's asset portfolio. Innovation teams will discover, formalize, and implement additional assets beyond what this book presents. The scalability of the methodology within the Five Verbs framework is one of its strengths.

Adopting documentation and becoming an asset factory might not be easy, but both are ruthlessly straightforward. In small steps, your company can evolve and, before you know it, become an asset factory—revolutionizing your language, habits, and culture.

An Asset Portfolio: Culture Disguised as a Template

The two best friends of execution are simplicity and transparency.
~ Chris McChesney (b. 1971), American author,
The 4 Disciplines of Execution.

The asset portfolio organizes assets in a few different ways. The portfolio groups some assets as process, people (supporting the Change Management function), and technology assets. Some assets are project-independent; some are project-specific. One group of assets captures the lens of the team, one captures the lens of individual team members, and another captures the intersection with outside organizations such as customers and competitors. The current state of the business is captured by a group of assets, as is the future state.

Figure 1 organizes the asset portfolio into three main groups to maximize customer and market centricity. Populating the assets in this order provides a natural path of least resistance and minimizes rework. The asset groupings explain how often each asset is used, that is, once for *project-specific* assets and cyclically (quarterly, monthly, or weekly) for *project-independent* assets.

In broad strokes, innovation teams create *project-independent* assets for decisions that influence multiple projects. They create *project-specific* assets for decisions unique to a project. And they create *technology* assets to govern the countless interactions humans have with the company's code and data.

Within these categories are subgroups of assets that fulfill the specific needs of the different asset groups. Independent of all projects, teams need *lens of the market* information to understand how business activity outside the company impacts innovation inside the company. Teams need *lens of the individual* assets so every employee can share progress, health, and ideas with their manager—vertically on an organization chart. Teams need *lens of the team* assets so that employees can share progress, health, and ideas—horizontally on an organization chart. *Current state* materials ensure teams comprehend the business before innovation work begins.

When working on specific projects, teams create *project management* assets to organize and report on project-wide activity. They create *process* assets to specify new stakeholder context and actions. Teams create *people* assets to proactively shepherd stakeholders through changes at an acceptable pace.

Technology assets govern every aspect of the company's technology actors (hardware and software). Teams need *design* assets properly sequenced to organize what the customer sees and how data is stored and retrieved for the customer. They use *build* assets so that code and data reside in the right places. And, finally, teams need *test* assets so that all stakeholders are confident that the processes and technology are working together as expected.

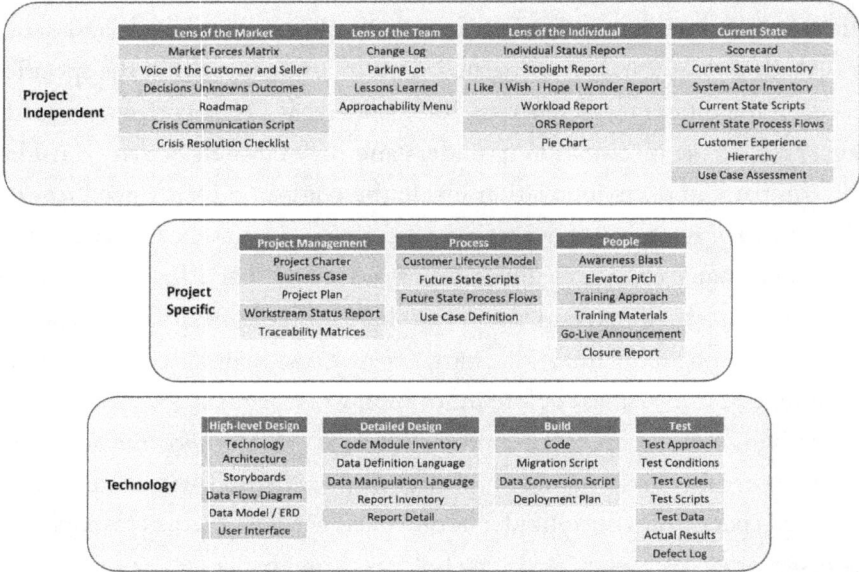

	Lens of the Market	Lens of the Team	Lens of the Individual	Current State
Project Independent	Market Forces Matrix	Change Log	Individual Status Report	Scorecard
	Voice of the Customer and Seller	Parking Lot	Stoplight Report	Current State Inventory
	Decisions Unknowns Outcomes	Lessons Learned	I Like I Wish I Hope I Wonder Report	System Actor Inventory
	Roadmap	Approachability Menu	Workload Report	Current State Scripts
	Crisis Communication Script		ORS Report	Current State Process Flows
	Crisis Resolution Checklist		Pie Chart	Customer Experience Hierarchy
				Use Case Assessment

	Project Management	Process	People
Project Specific	Project Charter	Customer Lifecycle Model	Awareness Blast
	Business Case	Future State Scripts	Elevator Pitch
	Project Plan	Future State Process Flows	Training Approach
	Workstream Status Report	Use Case Definition	Training Materials
	Traceability Matrices		Go-Live Announcement
			Closure Report

	High-level Design	Detailed Design	Build	Test
Technology	Technology Architecture	Code Module Inventory	Code	Test Approach
	Storyboards	Data Definition Language	Migration Script	Test Conditions
	Data Flow Diagram	Data Manipulation Language	Data Conversion Script	Test Cycles
	Data Model / ERD	Report Inventory	Deployment Plan	Test Scripts
	User Interface	Report Detail		Test Data
				Actual Results
				Defect Log

Figure 1: The Asset Portfolio

Figure 2 shows a high-level flow for a sampling of project-specific and technology assets. Process assets precede all technology assets, and people assets span the entire project's duration. Among the people assets, *awareness* and *desire* assets reside early in every project, *ability* and *reinforcement* assets reside at the end of every project, and *training* assets reside in the middle.

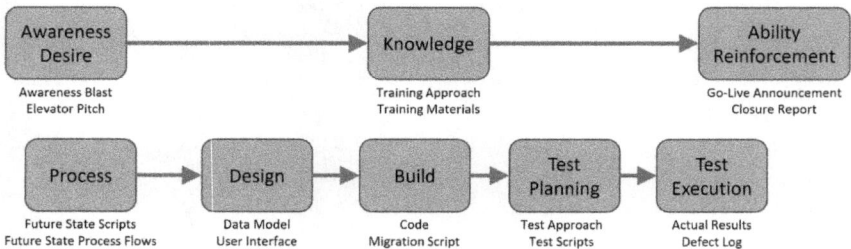

Awareness Desire	Knowledge	Ability Reinforcement
Awareness Blast Elevator Pitch	Training Approach Training Materials	Go-Live Announcement Closure Report

Process	Design	Build	Test Planning	Test Execution
Future State Scripts Future State Process Flows	Data Model User Interface	Code Migration Script	Test Approach Test Scripts	Actual Results Defect Log

Figure 2: High-Level Asset Flow with Representative Assets

16

The asset portfolio is valuable for both large and small teams. If your team is small, the assets don't become less valuable. Your size means you can adopt and complete them with less specialization and labor than large organizations and projects can. Small size means your small wins are more frequent.

But regardless of your company's size, no employee, no matter how senior, contributes to every asset or is exposed to the content in every asset. But it's valuable for every stakeholder to understand the nature and interdependencies of every asset to be competent in diagnosing problems on any project. A basic understanding of the asset portfolio equates to innovation literacy.

Ignorance of an asset is equivalent to saying that diverging expectations and disagreement on the information are no big deal. Diverging expectations are a big deal because they result in negative surprises. Willful ignorance sets people and projects up for failure. If you ignore an asset, a project won't automatically fail,[5] but success is more difficult. Dedication to these assets hinders the survival of systemic or systematic flaws. The assets reduce the laboriousness of information sharing, minimizing neglect, blind spots, and surprises. When employees can synchronize asset-building by systematically executing Five Verbs, what's left are only honest mistakes—sampling errors. Ruthlessness toward systemic errors and giving grace for honest mistakes shape a culture of discipline, empathy, value, and success—for people and projects.

Your organization might already use some form and combination of these assets. If that is the case, the sequence in which you continue building your asset portfolio is less important than progressing naturally and eventually building all of it.

---

5   The asset portfolio is so robust that downstream assets will force a correction.

How This Book Is Organized

How do you eat an elephant? One bite at a time.

~ Desmond Tutu (1931–2021), South African theologian
and human rights activist

This book defines and educates about the asset portfolio. Taking each of the three main groups of assets—project-independent, project-specific, and technology assets—in turn, each asset begins with a short textbook definition before sharing a complete description and guidance on how to populate the asset. Each *cross-project* asset has a recommended frequency; each *project-specific* asset identifies prerequisite and post-requisite assets. And, finally, every asset references, *in italics*, the culture traits related to the other two metaphors in the Elegance methodology—a factory and the arts (the emphasis of Volume One).

The topic of asset *sequence* warrants clarification. The reading sequence (for you) and execution sequence (for a project team) are similar, but not identical. As you might imagine, the execution sequence is more complicated than a table of contents or a diagram can show. The complicated relationships are why project-independent assets emphasize frequency and project-specific assets emphasize prerequisite and post-requisite assets.

In populating project-independent assets the first time, build assets in the order they appear in the book at a pace comfortable for your team. After that, review, revise, approve, and distribute the assets at the specified frequency (e.g., quarterly, monthly, or weekly) instead of strictly by sequence. Finally, stagger weekly assets over a month and stagger monthly assets over a quarter. Sequence assets how they appear in the book or how your team feels each asset impacts other assets. See Appendix C for a proposal of this asset staggering.

In populating project-specific and technology assets, innovation teams execute Five Verbs once per asset. Exceptions arise when the team learns something during the project to justify revisiting (reviewing and revising) the relevant assets. Each asset proposes its immediate upstream and downstream assets. An effective project planning tool visualizes a holistic view of these upstream and downstream relationships. Appendix C shows an example.

Finally, although the Project Plan asset is nestled in the middle of dozens of assets, its relationship to *all other* assets, project-specific and project-independent, gives it a unique importance. It is the boss, the traffic cop, the orchestra director. It prevents too much going on at once and too much waiting around. It encourages a team to set a visible, comfortable pace for working confidently and optimistically. It clarifies when and how employees contribute. It provides a sense of accomplishment and closure. Because it does all this using a finite and stable number of assets (approximately sixty) and verbs (exactly five), a junior employee can maintain the Project Plan as the project team completes the orchestrated asset portfolio. The Project Plan shapes and synchronizes a culture of discipline and empathy.

———•———

Over the past couple of decades, documentation has fallen out of favor for many innovation teams. These teams leave a poor legacy. Knowing what is worth keystrokes is a powerful defense to documentation skeptics. The right documentation improves the collaboration, value, and culture of teams. This book provides clarity and confidence to work on the right things at the right time, build your team's valuable asset portfolio, and establish a magnificent legacy.

PROJECT-INDEPENDENT ASSETS

*Give me six hours to chop down a tree and I will spend
the first four sharpening the axe.*

~ Abraham Lincoln, sixteenth president of the United States

When starting a project, many teams give little thought to the existing documents that will contribute to their success. A typical team focuses on future state documentation that is project specific. This section explains project-independent assets that optimize every project's speed, quality, and ease. These assets set up every project for success.

A team can document these assets without the pressure of project deadlines. Because the interdependencies of these assets differ from those of project-specific assets, the frequency of these assets is more relevant than their sequence. The optimal frequency is usually weekly or monthly.

Project-independent assets shape a culture of listening and awareness, and they allow companies to optimize globally. These assets capture the perceptions of stakeholders at the altitude of individuals, teams, entire organizations, and customer segments. Sharing the assets promotes an understanding of what's important to stakeholders. And the broad scope of the assets enables a team to consider the big picture before starting specific projects.

Successful businesses stay aware of relevant activity outside their walls, that is, what customers, competitors, and other market actors do and want. An innovation team documents certain information to give stakeholders confidence that the team is listening and prepared to receive new information from the outside. These assets are categorized as "lens of the market."

Good leaders are attentive to the perspectives of their teams. These perspectives include innovation ideas, lessons, issues, risks, and employee behavior. Advanced knowledge of what to document improves teamwork before and after the documentation itself. These assets are categorized as "lens of the team."

Good leaders are also attentive to the perspectives of individual employees. The foundation of this perspective is a basic status report. In addition, employees share compliments, concerns, and questions. They report on task health, operational burdens, and levels of chaos. They report whether their workload is too high, too low, or just right. These assets are categorized as "lens of the individual."

Innovation teams benefit from a healthy grasp of the current state, and the easiest way to achieve it is by having current state assets. If you lack documentation, get out of documentation debt by building the current state assets. They are a fantastic way to engage, educate, and integrate junior employees, which keeps costs low. Current state assets provide transparency that eases partitioning, prioritizing, and pacing projects. Building documentation is unpopular because of its upfront cost. But *using* documentation is popular since it reduces ongoing marginal cost in the form of laboriousness.

Many businesses have two other project-independent topics: products and solutions. Neither a product catalog nor a solution inventory fits into the asset portfolio. This is because an innovation methodology can't standardize them. Products and solutions should serve customer experiences, not dictate them. Their information resides *inside* process and data-oriented assets. And the simplest information about these topics is a kind of database query. For these reasons, products and solutions don't play a formal role in the Elegance methodology.

Project-independent assets shape the culture of individual projects. Executing Five Verbs (draft, review, revise, approve, distribute) ensures work is durable, not disposable. Five Verbs helps teams avoid meeting gridlock, email overload, and Verb Sprawl. Five Verbs instills discipline and empathy in collaboration. These assets set your team up for success by getting you out of documentation debt and firmly on to your innovation frontier.

Lens of the Market

The customer's perception is your reality.

~ Kate Zabriskie (b. 1970), USA-based shaper of business culture

The origin of your organization is not your team; it is not individual employees. Your organization's origin is outside your company. Its origin is the market in which it operates. This section describes six assets that capture the lens of your market: Market Forces Matrix, Voice of the Customer and Seller, Decisions Unknowns Outcomes, Roadmap, Crisis Communication Script, and Crisis Resolution Checklist.

A typical project team focuses on processes, people, and technology. Team members only sometimes see this information about external stakeholders, so they may not fully appreciate that the basis of compelling innovation work originates in information from their market.

Your market contains several stakeholder groups: customers, employees, partners, competitors, and your company. Valuable assets capture where these stakeholders have been, where they are today, and where you believe they're going. These assets capture long-term scenario planning, short-term activity, and crisis scenarios.

Skeptics might claim the information in these assets "belongs" to Sales, Human Resources, Strategy, or Competitive Intelligence teams. Who builds and maintains the assets isn't important. What's important is their existence and that they share the relevant information with project teams. This information is the origin of every project team's "why"—why teams build new customer experiences. Knowing its why sets a project team up for success. When a team doesn't know its why, it can waste time on projects that stakeholders don't value.

Lens of the market assets capture how the company competes and collaborates, how it understands outside information, what projects are next, and what to do in a crisis. These assets align information about the intersection of your company and outsiders. They contain information about the origin and purpose of the organization.

The assets with information at the innovator–customer intersection govern customer requests, education, and alignment. Using the metaphor of the arts, this collaboration resembles the artist–audience intersection.

Lens of the market assets are sequenced in the order a team would create them if they were new. Once they all exist, interdependencies are complex. What matters is the frequency of refreshing each asset (weekly, monthly, or quarterly).

Lens of the market assets shape your culture's discipline, forcing you to be *vigilant* about market *economics* and company *autonomy*.[6] They shape your culture's empathy, working toward your organization's desired *legacy* by being *flexible* and *blending* with other market participants.

Market Forces Matrix

The organization's "big picture" and strategic, market-level information such as revenue trends, cost trends, profitability goals, capability evolution, customer sentiment, and competitor activity.

---

6   Recall that the italicized terms in the last paragraph of sections are cross-references to Volume One of the Innovation Elegance series. These terms are culture traits found in the metaphors of a factory and the arts.

The Market Forces Matrix is the foundational asset that captures the company context to inform innovation decisions. It captures the organization's capabilities, current customer sentiment, competitor activity, and events in the value chain. It also captures financial information about revenue, costs, and profitability. The matrix embodies the company listening to and monitoring the market. It qualifies as an organizational dashboard at the highest level.

Innovators start with populating this asset because it exercises market awareness, that is, "meeting market participants where they are." The Market Forces Matrix pressures innovation teams to be customer-centric in simple ways using simple words. Starting elsewhere for innovation decisions is a sign of low customer centricity. Starting elsewhere is a sign that decision-makers listen more closely to their closest coworkers than customers and other market participants.

The matrix contains seven pieces of information spanning the company's finances, capabilities, and outside stakeholders. When a company's finances impact innovation, innovation originates as circumstances with some combination of revenue, cost, and profit. When external market participants influence innovation, the source of innovation is some combination of customers, competitors, and more distant players in the value chain. The asset should distinguish the actions, events, and capabilities of external stakeholders from those of your internal stakeholders.

The information in the matrix does not have to be comprehensive, and it shouldn't be. Trying to be comprehensive encroaches on the domain of non-innovation departments of a company, such as Pricing, Procurement, Finance, and Competitive Intelligence. Having this information in one place helps the team know the exact information that should factor into innovation decisions and avoid information overload. The matrix should only contain information that paints a picture of innovation considerations and decisions.

Here's an example of this asset for a fictional company called Outdoor Adventures, which organizes and hosts camping, hiking, and skating experiences and that is eager to innovate and grow.

Revenue	Capability	Customer	Competition	Value Chain
Market seems to be insensitive to price and tolerant of price increases	Terrain-neutral camping and hiking offerings are stable	Outdoor adventurers	Two players in adjacent markets exiting opens market to mountain adventurers (skiing) and urban adventurers (skates and bicycles)	Always sensitive to supplier costs for shoes, outerwear, and tents
Cost Fixed Cost: Rent cost is stable for next two years Variable Cost: Minimum wage likely rising to $15/hour				No disruptions seen at the moment.
Profitability Gross margins are 40% and stable				

Not having a Market Forces Matrix dooms a company to operating in a silo because it has poor self-awareness of its place in its market. The absence of a Market Forces Matrix suggests that goals related to revenue (with components price and quantity sold), cost (with components fixed cost and variable cost), and profit don't factor into innovation decisions. Maintaining this matrix formalizes a company's self-awareness of how its finances, capabilities, and outside stakeholders influence its innovation decisions.

Company information upstream of the Market Forces Matrix is outside the scope of the Elegance methodology and this book. That information includes quarterly financial results, publicly available announcements, and competitive intelligence. Downstream assets include Voice of the Customer and Seller, Scorecard, and Change Log.

To develop healthy habits around the Market Forces Matrix, assign and execute Five Verbs. First, innovation leaders assign a team member to *draft* the asset. Assigned team members conduct a quarterly meeting to *review* and *revise* the matrix. The end of every such session qualifies as *approving* it (or declaring it GETMO, i.e., "good enough to move on") for that quarter. Finally, a contributing employee *distributes* the matrix to relevant noncontributing stakeholders.

The Market Forces Matrix shapes your culture's discipline in a few ways. It shares your company's *economics* that influence innovation, the *variability* of capabilities, customers and competitors, and boundaries with outside companies (i.e., *autonomy*). The matrix shapes your culture's empathy by emphasizing the "artist–audience" intersection and maintaining *self-sufficiency*, *stewardship*, and *harmony* with other market participants. An *inclusive* "innovator–customer" intersection increases the size of the customer base and the number of profitable opportunities.

Voice of the Customer and Seller

Inventory of customers' and sellers' recommendations
for an organization's innovation decisions

Innovation starts with the customer. Without a customer, there is no business or innovation. The seed of innovation is empathy for customers. The concept of Voice of the Customer (VOC) formalizes this idea. To uncover ideas, problems, and additional revenue streams, a robust, customer-centric asset portfolio includes VOC.

VOC takes form in market research. Overt research techniques include surveys, interviews, and focus groups. Covert information collection uses sensors, internet cookies, and Internet of Things (IoT) devices. The field of VOC is large, and countless publicly available sources contain expert information. Savvy VOC work is not limited to current customers—former customers, prospective customers, and the competition's customers should all factor into VOC analysis.

Yet a commonly neglected voice is that of the salesperson. Typically, salespeople are the employees closest to the customer. Sellers are uniquely incentivized to serve the customer because they work where the company intersects with the customer. Instead of documenting solely Voice of the Customer, document this intersection—document the Voice of the Customer *and* Seller (VOCS).

Content in this asset doesn't guarantee approval. It guarantees that you are listening and considering ideas. Bad ideas appear and must be heard because they are often bridges to good ideas. When redundant ideas arise, the VOCS asset shows that your company listens to and considers all ideas. The asset reduces the time to process, explain, and defend decisions related to customers' and sellers' ideas.

The following page illustrates a template for the VOCS asset. To populate, give every idea a number and a title. Capture the date you recorded the idea. A problem/opportunity statement conveys what's wrong, and a future state description describes what's "better." Note the author, their inspiration, and the perceived impact (high, medium, or low) to understand the origin and value of the idea. And cite what is worth measuring about each idea. If the idea "moves a needle," label the needle.

Without a VOCS asset, ideas lounge, loiter, scattered in the minds of your customers and sellers. This slows and prevents your company from acting on their ideas. The absence of the VOCS asset hints that your company doesn't attentively listen to its customers or sellers. Poor attentiveness to customers and sellers cripples revenue. Populating the asset captures the best ideas at the intersection of your company's revenue producers.

#	Title	Date Suggested	Problem / Opportunity Statement	Future State Description	Author	Inspiration	Impact H M L	Relevant Metrics

The asset reduces the cost of sharing, refining, and mobilizing your next lucrative revenue streams.

Upstream assets include the I Like I Wish I Hope I Wonder report from customers and sellers. Downstream assets include the Scorecard, Change Log, and Parking Lot. Aim to revisit your VOCS monthly. VOCS information is typically more detailed than customer information in the Market Forces Matrix, and updates to both assets often coincide.

Healthy habits for the VOCS asset assign and execute Five Verbs. Junior salespeople[7] are ideal employees to *draft* the asset. Each month, diligent sales teams *review* and *revise* it. Sales leaders *approve* it (or declare it GETMO), and a junior salesperson *distributes* it to relevant noncontributing stakeholders.

The VOCS asset shapes your culture's discipline by *speeding* a team to process, filter, and approve high-*quality* ideas. The asset shapes your culture's empathy, too. It embraces *messiness* because many ideas expose bad products or services. It *sets the table* for customers and sellers to *co-create* ideas for additional value and revenue. The VOCS asset proves your company *listens* to its customers and sellers.

Decisions Unknowns Outcomes

Framework to illustrate future scenarios, probabilities, and preferred outcomes.

Crystal balls to see the future do not exist. And if you had one, your coworkers would likely have one too. The Decisions Unknowns Outcomes (DUO) asset captures your team's collective, hypothetical crystal ball. This framework helps your team collaborate on what it *wants* to happen and *believes* will happen so they optimize decisions and outcomes. This diagram

---

7   Junior salespeople and front-line associates are often hidden gold mines of innovation ideas. Assignments among Five Verbs improve the visibility and clarity of their ideas.

shows how decisions and outcomes can independently be good or bad and shows the label for each scenario.

Decision versus Outcomes[8]

Outcome Quality Decision Quality	Good	Bad
Good	Earned Reward	Bad Luck
Bad	Dumb Luck	Just Desserts

There is a high correlation between good decisions and good outcomes (earned reward), and more often than not, bad decisions lead to bad outcomes (just desserts). But outcome and decision quality do not always match. For example, being punctual (good decision) and perishing in an ill-fated plane crash (bad outcome) is bad luck, whereas being late and missing the flight of an ill-fated plane crash is dumb luck.

If a team isn't facing a scenario of risk, remorse, or regret, the DUO asset isn't valuable. However, if your team faces big decisions that involve risk, probability, and a large range of desirable and undesirable outcomes, this asset clarifies stakeholder preferences and personal risk profiles. The asset is valuable when team members have differences in perceived level of risk and personal risk tolerance and when they value different outcomes. The asset helps colleagues understand divergent biases and recommendations across a team. With it, the team can explore convergence and actions that improve probabilities and outcomes. The exploration's goals are better decisions *and* better outcomes.

---

8   Framework credit to Annie Duke, professional poker player and author of *Thinking in Bets*.

Without this asset, a team treats decisions, unknowns, and outcomes like informal bets. It's human nature to want to win a bet even if it means a colleague loses the bet, which sets up unhealthy competition within a team. In contrast, formalizing differences generates empathy within the team by exposing team members' blind spots.

The DUO framework has three ingredients:

- Sequence of future decisions and future events

- Perceptions of the probability of future outcomes (unknowns)

- The value placed on (or "preference for") those outcomes

Figure 3 is a generic example of the DUO framework. Its building blocks are as follows:

- Decision to choose Option A or Option B leading to either future Event F or Event G

- For Event F, the perceived probabilities of either Outcome W or X are perceived as equally likely (resembling a coin flip)

- For Event G, Outcome Z is perceived as much more likely than Outcome Y

- Of the four possible Outcomes, Y is the most preferred, Z is the least preferred, and W and X are only modestly preferred (but W preferred over X)

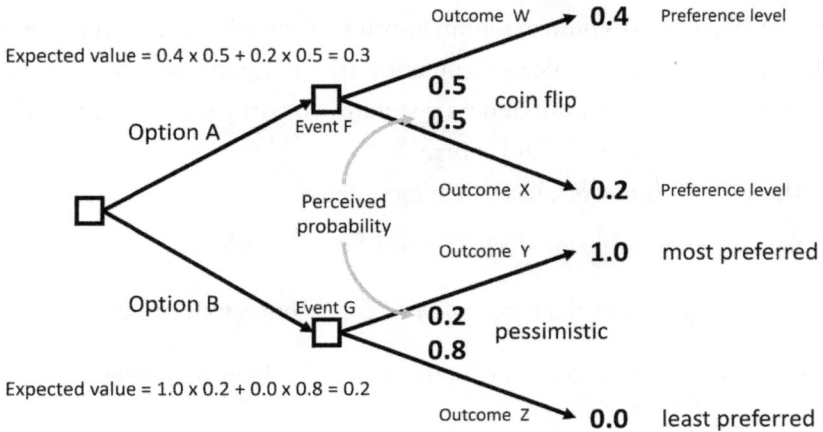

Figure 3. Decisions Unknowns Outcomes

Even though the most desired outcome (Y) is only possible by choosing Option B, Option A's expected value (0.3) is higher than Option B's expected value (0.2) (because of the high probability of the undesirable Outcome Z), so option A is the optimal decision.

Here are steps to build your decision tree:

1. Write three different simple lists (not yet in a decision tree format).
    a. Expected future decisions (in your control)
    b. Expected future events (not in your control)
    c. All possible outcomes

2. Arrange these lists chronologically into a decision tree that reflects:
    a. Branches that reflect alternating decisions (in your control) with events (not in your control)
    b. Endpoints at the right that represent the expected outcomes

3. Assign values (personal preferences) between 0.0 and 1.0 (tiered by tenths of one point) for each endpoint:
    a. 1.0 = your most preferred outcome
    b. 0.0 = your least preferred outcome

4. Assign probability at each event node (expressed as a decimal between 0.0 and 1.0)

5. From right to left, calculate the value (aka "expected preference") of each branch stemming from each decision node.

6. On the basis of your perception of probabilities and preference values, select your optimal path.

Even for a team of two, this methodology creates valuable insights and can convert personality conflict into task conflict. It helps users build understanding and see the world differently.

Figures 4 and 5 are examples of the framework for the fictional company Outdoor Adventures (OA).

OA serves the customer experiences of camping and hiking. OA's customer base and retail locations are more city-oriented than rural. OA's leaders learn that two competitors are going out of business. One competitor serves urban adventures such as roller skating. The other competitor organizes mountain adventures such as skiing. OA's leadership is in consensus about adding one customer experience at a time but disagrees on which customer experience to offer first. Some leaders recommend skiing. Other leaders recommend skating. The "skiers" see the scenario like this:

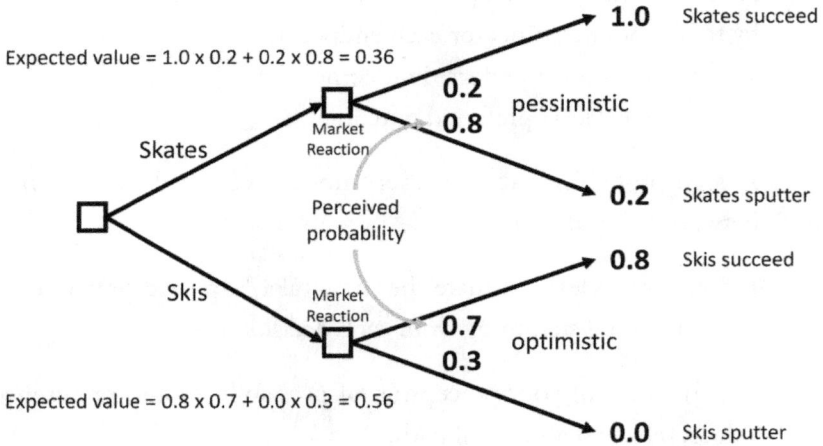

Figure 4. DUO from Skiers' point-of-view.

Skaters see their scenario like this:

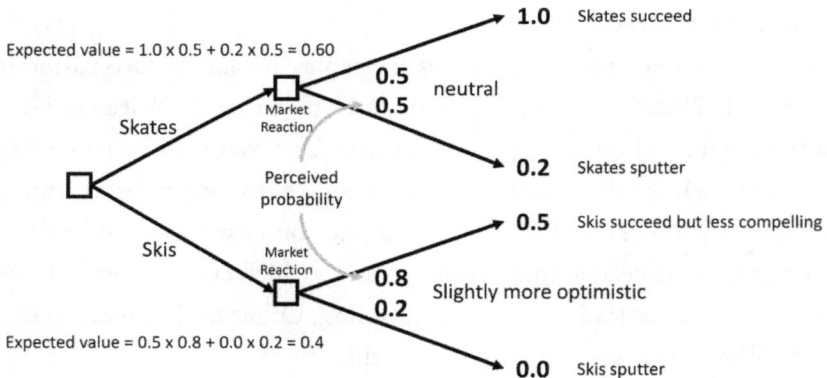

Figure 5. DUO from Skaters' point-of-view.

You can draw a few conclusions from these diagrams. Both groups see the highest profit potential in skates because cities have a large customer pool that is more aware of OA as a brand. Skaters see the expected profits of skis as less compelling than skiers do but are more confident than skiers in skis' reliability of profit. The big difference in the model is that skaters see a recent surge in the popularity of activities considered "retro," like roller skating. The skiers don't believe this translates to skating. Skaters are more optimistic than skiers about skates' prospects for large profits.

At a minimum, each group understands better how the other group answers certain questions differently. For example, do they see the same quantity and description of outcomes? Do they value those outcomes differently? Do they see the same probabilities? And do they see a different decision for the team? A richer collaboration entails team members influencing each other and redrawing this asset to reflect different decisions, unknowns, probabilities, and outcomes.

The asset serves as a sanity test for a person's intuition. When expected values of options have minimal differences, the model likely reinforces that a decision is a "tough call" (difficult). When expected values are far apart, the model likely reinforces why a decision is a "slam dunk" (easy). A small amount of simple yet formal math helps the team make decisions, although it doesn't completely replace informal human judgment. If the model yields expected values that contradict your intuition, revisit the expected preference levels and perceived probabilities.

The absence of the DUO asset undermines common understanding and perceptions of events' sequence, values, and probabilities. Without the DUO asset, teams tolerate poor awareness of an organization's diverging risk profiles and motivations. This asset aims for shared understandings, better individual and group decisions, aligned incentives, and better outcomes. You can show this asset to your customers and stakeholders and have them share their perspectives, too.

When investment decisions feel messy and overwhelming, this framework combats the messiness by simplifying the variables and painting an elegant picture of your innovation's future.

Upstream assets include the Market Forces Matrix, Change Log, and Parking Lot. Downstream assets include the Roadmap and Scorecard. A typical frequency to update DUO is monthly, depending on how often stakeholders change their minds about event sequences, values, and probability.

As with the other lens of the market assets, the DUO asset should be put through the Five Verbs process. Any innovation decision-maker can *draft* the asset. As events, perceptions, and values change, their peers should *review* and *revise* the asset. A manager of the contributors should *approve* a reconciled version of the asset, yielding a decision to pursue. Contributing stakeholders *distribute* the result to other stakeholders by sharing the actual asset or just the conclusion.

The transparency of the DUO asset shapes your culture's discipline by increasing *confidence* in a team's decisions. This confidence improves *speed* by reducing the number of hasty and procrastinated decisions. Knowing what unknowns to monitor improves *vigilance*. The asset shapes your culture's empathy by encouraging *authenticity* in expressing preferences, a *balance* in perspectives, and *positivity* in decisions that optimize more globally than locally.

Roadmap

> A visual display of current and future projects the team aims to execute in the next three to twelve months.

A healthy innovation team formalizes its in-flight and near-future projects.[9] The asset to convey this information is a Roadmap. A Roadmap contains a plain list of projects, a calendar, and notations showing each project's approximate start and completion dates.

A Roadmap captures the "what" and "when" of innovation priorities along with their intensity[10] and duration. Other assets (explained *later* in this

---

9   The word team here is used very loosely to mean any combination of people who would contribute to an asset such as a Roadmap, not a formalized department or reporting structure. The boundaries for the components of a Roadmap are subjective and unique to every organization.

10   Intensity is reflected in the quantity of simultaneous projects. Three simultaneous projects might qualify as low intensity. Fifteen simultaneous projects might qualify as high intensity.

book even though some are *upstream* in project execution) capture evaluations and the "why" of these topics. Contributors list the names of projects, when they start or have started, and when each is expected to finish. At a glance, consumers of the asset can learn these names and dates. A Roadmap's simplicity makes it easy to change the priority, intensity, and duration of projects, but its transparency discourages impulsive changes. A healthy team revises a Roadmap every month to expose any information changes.

A great Roadmap only includes projects, that is, one-time-only workstreams with expected start and completion dates (see the example below for Outdoor Adventures) that have been approved via the Change Log (discussed later). A Roadmap excludes *operations*, the continuous or repeated workstreams such as audits or taxes that have predictable frequencies of monthly, quarterly, or annually. A Roadmap shows projects that *impact* operations, but a Roadmap excludes an inventory of *actual operations*. (The section "Current State," later in the book, explains how to document operations.)

An innovation Roadmap shows an organization's innovation priorities. Projects with early start dates are high priorities. Projects with later start dates are lower priorities. If you record project priorities in a text-only format with the letters *H*, *M*, and *L* (representing high, medium, and low), a common result is that everything is a high priority. This means that nothing is a high priority. A visual format like the monthly calendar below forces tough choices about priority.

The visual format of the Roadmap governs project density. In the following example, OA has one project in flight in December. A few months later, the company has more than one project in flight at a time. A Roadmap helps keep a sensible number of projects (as subjective as it might be) in flight and avoids having too much, too little, or too many fluctuations in what's going on. Too few projects are a sign of underinvestment and planning neglect. Too many projects are a sign of saturation, gridlock, or idle work. Fluctuations—"lumpiness"—typically require intense hiring or firing sprees that pose separate relationship and reputational risks.

This asset visualizes expected completion dates. Projects with long durations, because of their high costs and vulnerability to obsolescence and changing priorities, can be scrutinized. What a Roadmap easily exposes is *changes* in duration. If a project increases in duration on a Roadmap from one month to the next, the project has delays or else languishes. Because lengthy projects draw additional risk and scrutiny, a Roadmap encourages teams to keep projects modest in scope and duration.

Figure 6 is an example Roadmap for Outdoor Adventures. It shows a handful of customer experiences and products the company plans to add over a fifteen-month window.

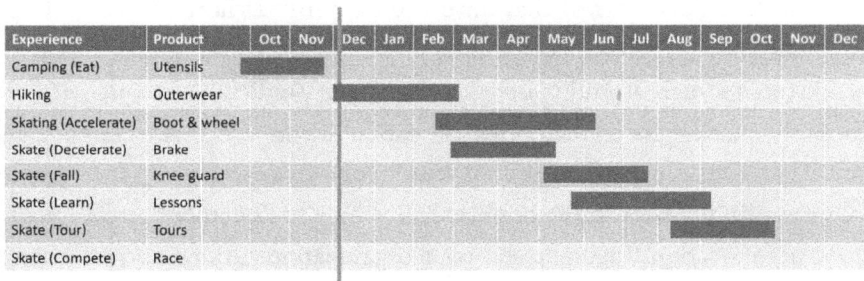

Experience	Product	Oct	Nov	Dec	Jan	Feb	Mar	Apr	May	Jun	Jul	Aug	Sep	Oct	Nov	Dec
Camping (Eat)	Utensils															
Hiking	Outerwear															
Skating (Accelerate)	Boot & wheel															
Skate (Decelerate)	Brake															
Skate (Fall)	Knee guard															
Skate (Learn)	Lessons															
Skate (Tour)	Tours															
Skate (Compete)	Race															

Figure 6. Roadmap.

Project titles start appearing upstream of the Roadmap in the Change Log. Those titles may or may not be meaningful to customers or to those primary stakeholders who are affected. When your team transfers approved projects—individually or bundled—to the Roadmap, try to give them a customer-centric title. One reason is that you might share your Roadmap with customers to set schedule expectations. Another reason is that repeating customer-centric language has a positive domino effect during the project, improving engagement, morale, and adoption.

The pending role of Five Verbs can be seen in the incomplete row for Skate (Compete). Once the Roadmap is *drafted* and project titles added, assigned team members conduct a monthly meeting to *review* and *revise* the Roadmap. This includes setting approximate time spans for new and existing projects. Rough convergence at the end of the monthly meeting constitutes *approval,* and one of the contributors *distributes* the Roadmap to relevant noncontributing stakeholders.

A Roadmap's strength is its simplicity. Other assets govern the "why" of project priorities and contain detailed rationale about priority and duration. A Roadmap requires precision in the titles and intensity of projects but not for their schedules. Because poor alignment on a Roadmap's information results in major surprises, a Roadmap provides a simple format to minimize major surprises. Monthly revisions to a Roadmap align big-picture expectations such as a project title, the approximate schedule, and the expected delays.

Skeptics of the Roadmap claim they cannot envision, prioritize, or stabilize the following year's projects. This claim equates to surrendering to VUCA (volatility, uncertainty, complexity, and ambiguity), as if their business is too volatile, uncertain, complex, and ambiguous to plot a course. Like every asset in your portfolio, a Roadmap is not rigid. Its simplicity accommodates changes easily. Skeptics might feel the effort and vulnerability required to look into the future is too great, but a Roadmap forces a team to be proactive and to plan for its future. It encourages a company to be responsive, not reactive. It encourages a company to explore opportunities with customers more ambitiously.

A different kind of skeptic doesn't want to expose the flooding, gridlock, and chaos of dozens of in-flight projects. A roadmap is the perfect asset to ease traffic jams, reduce change saturation, and restore a sustainable intensity of innovation work. A roadmap exposes languishing projects so they can get the right attention toward completion or cancellation.

The Change Log and Use Case Assessment are upstream assets that drive project prioritization. The Project Plan and Workstream Status Reports are upstream assets that show duration. Workload Reports are assets that influence intensity. The Roadmap's downstream asset is the Project Charter.

Upstream assets drive monthly changes to the Roadmap. Newly approved items on the Change Log justify adding new projects to the Roadmap. Effort and impact (details on the Change Log) are starting points to determine priority (start date on the Roadmap), and tough choices can overrule what's on paper. The Project Plan and Workstream Status Reports influence start and completion dates. Workload Reports influence project intensity. It's healthy to keep just-completed projects on the Roadmap for one or two months as a celebration but be sure to remove projects once they are sufficiently in the past.

A Roadmap shapes a culture's discipline. Project start dates represent innovation initiatives that are economically minded and prioritized. Capping intensity and monitoring duration allow for projects to move at respectable *speeds*. A Roadmap shapes a culture's empathy by encouraging *paced* innovation, discouraging *fatigue*, and *setting the table* for customer-centric work for months into the future.

Crisis Communication Script

> Template of what information to share
> on a regular cadence during a crisis.

A Yiddish proverb reminds us that "Man plans and God laughs." As well-planned as an innovation team's work might be, crises will occur. Good leaders share information with stakeholders in clear, straightforward terms. Stakeholders have common questions, and leaders can answer these questions with a Crisis Communication Script.

A conventional executive might be willing to under-communicate or outright hide a crisis. A bad leader might craft messages that conceal the severity of a situation. A disciplined, empathetic leader instead crafts messages with accuracy and transparency. In this era of recordings and social media, poor communication can have terrible results for leaders and their stakeholders.

A good leader shows up, especially in a crisis. Good communication during an emergency can save lives, property, and reputations. A responsible leader does not have to be long-winded, nor do they have to bare their soul. Overcommunicating muddles stakeholder understanding and dilutes focus from optimal reactions. Extraneous information—often called a "red her-ring"—hurts a leader's credibility. Especially when a crisis is new and many details are unknown, good leaders are upfront about what they don't know.

The Crisis Communication Script answers the following questions:

1. What do we currently know?

2. What do we not yet know?

3. What are we doing to know more?

4. What teams or junior employees are pursuing a solution?

5. What senior employees hold accountability and authority?

6. When will leadership be back in touch with more information?

Skeptics of a script lack discipline and empathy for their stakeholders who might benefit from knowing this information. Skeptics both over and undercommunicate. They let rumors spread until they lose control of the narrative. Good leaders avoid silos but still shape tight messages that work for different communication channels and for all relevant audiences.

After one recent highly criticized business crisis, Tennessee-based author, speaker, and publisher Michael Hyatt recommended leaders claim "extreme ownership" with words like these: "Something went terribly wrong here. The full responsibility lies with my organization—specifically with me. We will do whatever it takes to make things right and learn from this experience so that it never happens again." Another high-profile executive who recently suffered a crisis said, "Crises need to stay front and center. We constantly want to be reminded of how things can go so wrong so quickly."

During hard times, leaders' values are tested. The words of a poor or false leader might merely be a hollow expression of concern. A crisis script prevents someone in power from abandoning stated values or abdicating accountability. For good-faith leaders, a script eases adherence to values such as accountability, transparency, and safety during hard times. For bad-faith leaders, a script helps those around them (because of transparency in the Five Verbs) prevent divergence from what best serves stakeholders.

The frequency of crisis messaging depends on the expected duration of the problem. A crisis lasting years warrants messaging every month. A crisis lasting months warrants messaging every week. A crisis lasting days warrants messaging every day.

Five Verbs optimizes the message content. A senior Operations employee *drafts* answers to the questions and consults stakeholders close to the crisis to *review* and *revise* the answers. The senior employee *approves* and *distributes* the message via communication channels (e.g., audio, video, email) that work for stakeholders.

A Crisis Communication Script advocates a culture of discipline. It works to recover the *quality* of the stakeholder experience. Regular messaging shows *vigilance*. Answering the six main questions minimizes *variability*. The script shapes your culture's empathy by showing *self-awareness* that a *failure* occurred. The script aims to be *transparent*, keep stakeholders *calm*, and prevent information from getting out of *control*. A script enhances *trust* in leadership to keep stakeholders *safe*.

Crisis Resolution Checklist

The sequence of assets to review and revise during a crisis.

A crisis seems like an understandable excuse to cut corners or flail about, but a Crisis Resolution Checklist (CRC) helps innovation teams avoid desperation and resolve the situation. The CRC guides you to do the right things in the right order without reinventing the wheel. A crisis-related proverb says, "The time to fix your roof is when the sun is shining." Creating a CRC resembles roof repair while not embroiled in crisis.

You might already have in place the traditional disciplines surrounding crisis management: BCP (Business Continuity Planning) and DR

(Disaster Recovery). Countless resources exist to educate you on those conventional methods. The Elegance methodology adds the CRC to this tool set. The CRC, assuming you have certain assets already in your asset portfolio, governs crisis resolution by identifying which changes need to be made to the asset portfolio. On your first reading of this book, assets set in *italics* in this section might not yet be familiar to you. Return to this section after a complete reading of the book or after your team builds these assets to understand the logic of the checklist.

When crisis resolution involves changes to processes, people, *and* technology, it amounts to an entire project, and there's no need for special governance. The CRC applies when the resolution is isolated to people and process assets. The checklist proposes that employees pause specific in-flight work and dedicate themselves to defining and executing a new or revised process.

The first decision is whether to pause in-flight *innovation*. You might pause a *portion* of a project, such as postponing a training event but allowing testing to continue. Alternatively, you might pause an entire project, that is, all its in-flight work. Both changes warrant changes to the *Project Plan* and the *Roadmap*.

The second decision is whether to pause in-flight *operations*. At a high altitude, you might pause a process in its entirety. This translates to revising your *Customer Experience Hierarchy*. At a low altitude, you might reassign some specific work. This translates to revising some *process flows*, effectively creating *Future State Process Flows*.

The goal of pausing in-flight innovation and operations is to liberate employees from certain work so they can formulate new and different processes that resolve the crisis (which resembles a new project).

At a high altitude, the third consideration is building a *Project Charter* for the crisis project. At a low altitude, this new project impacts existing processes and possibly warrants new processes—*Future State Scripts* and *Future State Process Flows*. Dedicating resources to define new processes is easier when you have *Current State Scripts* and *Current State Process Flows* to review, revise, approve, and distribute. In an emergency, build these assets within minutes or hours.

The fourth consideration is implementing the people dimension of the change, that is, training and mobilizing the new process to resolve the crisis. At a high altitude, the project can leverage or revise an existing *Training Approach*. At a low altitude, the crisis team must build and share *training materials*.[11]

The fifth and final step is to broadcast the new actions that resolve the crisis (temporarily or permanently). This can be translated to a *Go-Live Announcement* that can be delivered naturally in the normal rhythm of a team's *Crisis Communication Script*.

For the fictional company Outdoor Adventures, an example crisis involves discovering that a particular tent model fails, and it's unclear whether the problem is the product or a procedure. The first step is to consider stopping in-flight innovation, which impacts the Roadmap and Project Plans. The second step is to pause camping operations, which affects the Customer Experience Hierarchy and Current State Process Flows. The third step is to start a project called "Restore Camper Safety," which involves a Project Charter, Future State Process Flows, and engaging with the tent manufacturer. The fourth step is to build and distribute training materials. The final step is to build and distribute a Go-Live Announcement.

Note that the CRC lacks a sense of pacing. Where this checklist cuts corners is at the Project Plan. In a crisis, avoid fine-tuning timing or assignments in a Project Plan. Conduct large meetings (expensive and wasteful outside a crisis) to minimize blind spots and misalignment and, as a default, aim for this three-day schedule: On the first day of the crisis, pause everything you see fit. On the second day, complete the new project and deliver training. On the third day, execute the new processes. "Nobody leaves the room" until the crisis team executes Five Verbs for each relevant asset. Extend this schedule so it's feasible for your organization. Calibrate the

---

11   Training materials are a vital asset in every project, so the Asset Portfolio and Project Plan includes them. Because they are rarely neglected and impossible to standardize, this book doesn't include a section for the asset.

frequency of your Crisis Communication Scripts to fit what's realistic with your resolution checklist.

Skeptics of the CRC surrender to volatility, uncertainty, complexity, and ambiguity. Typically, they don't have these assets available and, likely, have never been crisis response team members. They lack the discipline and empathy to respond adequately when finances, reputations, and people's physical and emotional safety are at risk. Skeptics should acknowledge that a crisis is a high-pressure Q&A. The team performs better when members know the questions before they need to answer them. The questions *are* the CRC.

The CRC encourages your team to pivot within minutes of a crisis appearing. Charles Darwin (the father of the theory of evolution) wrote, "It is not the strongest of the species who survive, nor the most intelligent, but the one most responsive to change." This checklist galvanizes your team to respond to disruption.

A CRC shapes your culture's discipline in a few different ways. The checklist *speeds* the reassignment of employees to the crisis resolution project. A crisis is often an *economic* disaster, and the checklist enables you to react accordingly to salvage revenue streams and minimize legal or regulatory costs. The checklist keeps *variability* low so that innovation teams focus on the proper assets to minimize damage.

The checklist shapes your culture's empathy as well. It improves a team's *resilience* to negative surprises, which prevents a negative *legacy* from forming for an organization or a single leader. It also quickly shifts the organization out of crisis mode and into a state of "good enough to move on" (GETMO).

———

Among the project-independent assets in your asset portfolio, the lens of the market assets sit at your company's boundary with the market. Everything your company innovates is for your customers. Information going in and out of your company drives not only your innovation team but also everything else inside your company. Five Verbs governs the agreement factory at the boundaries of your company.

Lens of the Team

None of us is as smart as all of us.

~Ken Blanchard (b. 1939), American motivational speaker, author, *The One Minute Manager*

Four assets capture team-wide, project-independent ideas, concerns, and mentoring: Change Log, Parking Lot, Lessons Learned, and Approachability Menu. Although these assets are not exposed to outsiders, their transparency provides vital information sharing within an innovation team.

These assets require generous, authentic, and vulnerable reflection from team members about issues, risks, individual performance, team performance, and about the most significant points of pain for all stakeholders. The goals include minimizing low-value events and counterproductive behavior while maximizing high-value events and collaborative behavior.

A typical team has a high tolerance for personality conflict, favoritism, and egos. Those three culture traits optimize locally and create silos. They hurt a team's long-term health and value. But, similar to how individual employees contribute value to their managers, a healthy team collaborates and maximizes their collective value to their leadership. A healthy team has high expectations for pragmatism and humility.

Skeptics of team introspection are self-centric. They aren't curious about their colleagues' ideas, and they don't trust their colleagues' judgment. Healthy mentoring among team members doesn't interest them. Skeptics diminish the value and purpose of having a team. The lens of the team assets shepherd a team toward being smarter than its individual members.

A core benefit of a team is that it provides a venue for the interaction of different ideas—ideas being a bridge to better ideas. The interaction leads to a team maximizing its value by replacing personality conflict, favoritism, and egos with task conflict, career security, and optimizing the big picture. Lens of the team assets move employees' attention off themselves toward learning from their team members and minimizing counterproductive patterns. These assets shape a culture in which employees care about each other's points-of-view, morale, expertise, and value. These assets enable every team member to contribute and understand team-level topics.

Lens of the team assets shape your culture's discipline by improving the *quality* of innovation ideas and employee behavior. These assets shape your culture's empathy by inducing *listening*, *authenticity*, and *trust*.

Change Log

> A form of "suggestion box"—a list of current problems, the desired state, and priorities for change.

Healthy innovation teams capture ideas for innovation. A Change Log resembles a suggestion box in that it can capture innovation ideas from any stakeholder. The goals of using a Change Log are to receive stakeholders' ideas publicly, loosely estimate their value with prioritization in mind, and approve the most valuable, time-sensitive ideas.

Many companies get big innovation ideas from executives and outside consultants. These stakeholders are rarely the closest to the most significant problems, so they miss numerous ideas while promoting their own opinions. A Change Log is best maintained by employees most familiar with the biggest stakeholder problems. Instead of laboring to achieve employees' buy-in on projects, your organization can use a Change Log to make employees a primary source of project ideas.

Whereas a Roadmap captures the "what" of innovation priorities (project titles) and the "when" (start date of each project), the Change Log captures a first snapshot of the "why" in innovation ideas and priorities. Language in the Change Log might be most persuasive when authors know their ideas effectively compete among many innovation ideas.

The Change Log is unique among the assets in the asset portfolio because this is where exciting new ideas make their first formal appearance. Many innovation assets feel serious, but the Change Log is the right place to capture ideas that delight stakeholders. As the first formal proposal of an idea, precision in wording isn't essential. Top-of-mind descriptions suffice.

Many companies use a Change Log that records much of the same information as below, but variations exist. A template for the Change Log is shown on the next page.

Healthy habits for the Change Log assign and execute Five Verbs. Maintaining a Change Log involves adding new items, judging candidate ideas, and transferring approved items to the Roadmap.

Assigned team members *draft* the first version of a Change Log. Authors should not be restricted from adding (*drafting*) new ideas to the Change Log. Invite stakeholders to add to the log informally (as if it were an old-fashioned suggestion box). In the log, number each item and give each a title. Describe the problem, typically the "current state" and what the author wants the organization to "run from." Describe the "future state"—what the organization should "run to." Name a junior employee who can articulate the idea. If the junior person can designate a senior employee to serve as a champion, include that name, too. Give a tier (high, medium, or low) to the estimated effort and impact of the change and give new items the status "New."

#	Title	Problem / Opportunity Statement	Future State Description	Author	Champion	Effort H M L	Impact H M L	Status

Assigned team members (often loosely called the "Change Board") conduct meetings to *review* and *revise* the log—monthly is a reasonable default cadence. The primary goal of every meeting is to approve, reject, or defer each new idea (assuming the Change Board has enough information to make a decision). The easiest ideas to approve are low effort and high impact, which are often the most profitable to the company. The ideas most difficult to approve (or easiest to reject) are high effort and low impact.

If the board is indecisive about an idea, it changes the status to "Open" and tries to decide the following month. Rejected and deferred items quietly stay on the Change Log to minimize rework while making it easy to revisit an idea at a later time. Typical status labels are New, Open, Approved, Rejected, and Deferred. Every month, the Change Board does its best to align on decisions. Aligned or not, the end of every month's formal update of the Change Log represents *approving* it (or declaring it GETMO) for that month. A junior team member *distributes* the Change Log to relevant noncontributing stakeholders.

If your team wants to record more information for each item on the Change Log, consider these:

- Impacted Customer Experience(s)

- Category (e.g., Customer, Competition, People, Process, Technology)

- Cost of Delay (lost revenue or inflated cost per day or month)

- Funds Available (Y/N)?

- Date Requested

- Desired Completion Date

- Next Checkpoint Date (for deferred items)

- Relevant Metrics

Another distribution step is transferring the approved items from the Change Log to the Roadmap. The Change Board might approve an item as-is for a one-to-one transfer to the Roadmap. Alternatively, the board might split one idea into multiple projects or bundle multiple ideas into one project. Because of this, the titles of ideas on the Change Log might look very different from their corresponding project titles on the Roadmap.

The discipline of bundling is known as "Release Management." For example, if a team bundles thirty changes into three releases in 2029, the team can refer to these releases as 29.1, 29.2, and 29.3. Conversely, one item from the Change Log might impact multiple customer experiences and translate into multiple releases.

The Change Log is the first appearance of project titles, or at least *candidates* for project titles. Some titles on the Change Log never progress to the Roadmap; when approved, others change for various reasons in the vetting and bundling process.

Skeptics of a Change Log prefer to pursue innovation or disruption in less transparent ways. They want favoritism and their relationships to secure approval for pet projects or change ideas from a smaller pool of stakeholders. This reduces healthy competition for investment and guarantees that lower-value projects appear on the Roadmap.

A Change Log is a catchall. Playful, aspirational stakeholders call the Change Log "stakeholders' boundless aspirations" and their "universe of ambitions." The Change Log has many upstream assets, including the Market Forces Matrix, Voice of the Customer and Seller, Decisions Unknowns Outcomes, Parking Lot, Scorecard, Use Case Assessment, I Like I Wish I Hope I Wonder, and Actual Results. The Change Log's downstream asset is the Roadmap. The typical frequency to update is monthly.

A Change Log shapes your culture's discipline by approving a team's highest *quality*, most impactful, and most *profitable* changes. A Change Log governs *elasticity* since a backlog of ideas exposes *idle work*. A Change Log shapes your culture's empathy by capturing the universe of ambitions, *listening* to stakeholder ideas, capturing the *passion* in ideas, and pursuing *positive surprises* for stakeholders.

Parking Lot

> A place for barriers, risks, issues, and questions that
> a team wants to manage individually to closure.

Although every asset is a form of Q&A, topics arise that warrant temporary attention outside the cadence of Five Verbs. A team might categorize such a topic as a barrier, risk, issue, or question. Closure for each is, respectively, a bridge, mitigation, resolution, and answer. These topics justify an asset of their own—more than an issue or risk log, it's called a Parking Lot.

A typical team might handle such topics extremely informally—with a side conversation in a meeting or a short email exchange. Sometimes that's not a problem, but sometimes such informality creates blind spots for stakeholders not in the informal communication loop or causes them to make assumptions. Erring on the side of formality in communication of these topics fosters clarity and transparency and elevates their importance because Parking Lot items can have a higher profile than those embedded in the assets where the information ultimately belongs.

At its best, a Parking Lot is a vehicle that enables stakeholders to "see around corners" and raise topics that deserve special attention. Mark Langley, a recent president of the Project Management Institute (PMI), encourages teams to celebrate these items as gems because they can be addressed openly. Stakeholders should be encouraged to find negative surprises before the surprises find the team.

A high volume of items in the Parking Lot hints that a team is trying to answer questions in the wrong sequence. For example, it's premature for a team member to insist on talking about training when the project is collaborating on the Project Charter. It's late for a team member to second-guess process assets while the project is collaborating on training materials. However, Parking Lot items might contain topics that a team works on out of order; unless explicitly logged, the team might not give a topic adequate attention when the time arrives for the right asset.

Once a Parking Lot topic has closure, the information doesn't simply stay in the Parking Lot but should be transferred to another asset. Three types of information transfer are the most common. When a Parking Lot item pertains to an assignment or the project schedule, transfer the information to the *Project Plan*. Alternatively, an item can expose a blind spot about the current state, and closure justifies additions or changes to *current state assets*. Or an item can identify additional stakeholders, impacting the *Project Charter* and multiple *people assets* (explained in later sections).

It's best to record all four categories of topics in the same list. Even though it's common and understandable for team members to disagree on categorizing a topic as an issue, risk, question, or barrier, these nuances are not important enough to merit the team's time. A template for a Parking Lot is shown on the next page.

In your Parking Lot, number each item, give it a title, cite the originator, and provide a detailed description. Approximate a priority so the team addresses items in the right sequence. Propose one or more stakeholders who can likely help. Record the status as open or closed or specify when the team should discuss it again. Reserve a space for explaining closure, that is, answering the question, resolving the issue, or mitigating the risk. And, finally, list the assets that should reflect this information permanently.

Skeptics of a Parking Lot either dislike giving transparency to such topics or they are too timid to speak up. Parking Lot topics often contain negativity or bad news. It's human nature to avoid these in the open, but it's a bigger crime to suppress solutions that prevent bigger problems.

Occasionally, a Parking Lot item reflects a team member's performance or behavioral problem, for example, hinting "Tracy is an issue" or "Manu is a risk." The Parking Lot is a place for concerns—not coaching. Other assets, such as the *Approachability Menu* and *Lessons Learned* template, are the right assets to address sensitive, performance-related topics.

#	Topic /Title	Author / Originator	Long Description	Priority H M L	Point Person /Need Help From	Open / Closed / Discuss Next On	Closure Explanation	Deliverable(s) Impacted
1								
2								
3								

Common upstream assets of a Parking Lot are the Project Charter and the Workstream Status Report, but more commonly, these items appear at any time and are independent of any specific asset. As mentioned above, the most common downstream assets are the Project Plan, current state assets, and people assets.

Healthy habits for the Parking Lot assign and execute Five Verbs. Assigned team members *draft* the first version of the Parking Lot. Any stakeholder can add items at any time. Assigned team members conduct a weekly meeting to *review* and *revise* it. In these meetings, prioritize items assigned "open" status, assigned an early date to "Discuss Next," and assigned high priority. Every meeting assumes that items marked "Closed" do not need attention, and ideally, the tool storing the Parking Lot can make closed items invisible in the meeting. Assignments for the Parking Lot change more frequently than other project-independent assets. The end of every meeting qualifies as *approving* the asset (or declaring it GETMO). Every team has a tool and mechanism to *distribute* the Parking Lot asset to relevant noncontributing stakeholders.

The Parking Lot shapes your culture's discipline by encouraging *vigilance* among stakeholders. Closure of the items prevents *quality* problems. A Parking Lot shapes your culture's empathy by fostering *safety* in the workplace, acknowledging *messiness* in teamwork, and *listening* to uncover blind spots.

Lessons Learned

A team's reflection on the most instrumental dimensions of team health and performance.

Lessons Learned is a common exercise—typically one with low rigor. But a team reflecting with high rigor can keep discipline and empathy levels high.

Because innovation and teamwork can be so complex, you might think a million things could go wrong. But that's not true. There are patterns in the problems. You can capture most problems by focusing on approximately a dozen culture traits. Continually reinforcing and calibrating these traits keep your culture disciplined and empathetic.

The conventional Lessons Learned exercise (or "retrospective" as Agilists call it) asks two questions: "What did we do well?" and "What would we do differently?" The low rigor of these two questions doesn't allow a team to scrutinize the culture traits that most commonly determine success or failure.

The culture traits in the template below capture the patterns in problems. Prompting an innovation team monthly with these traits maximizes the chance the team will discover behaviors and habits to fix. Even without formal responses, repeating and reinforcing the culture traits in the template can influence and improve behavior. In the early months of using this template, the goal is to minimize damaging habits and behaviors. After improving the worst of those, the goal is to maximize and reinforce valuable habits and behaviors, pursue an ambitiously collaborative culture, and maintain advantages that attract employees, customers and partners.

The template prompts team members to finish any or all four sentences for each trait.

"I like ____" contains positive reinforcement for behavior from the present, past, or future.

"I wish ____" refers to something from the past that the employee would like to be different.

"I hope ____" refers to something in the future that the employee would like to be different.[12]

"I wonder ____" reflects a neutral stance or a politically sensitive question about something in the past, present, or future.

The Lessons Learned template is on the following page. Every culture trait is discussed in more depth in Appendix D.

---

12 I Wish and I Hope occupy one column to reduce duplication of a single idea and soothe finger-pointing of backward- or forward-looking remarks.

Area	I Like	I Wish	I Hope	I Wonder
Safety, Inclusivity, Belonging			
Transparency			
Simple and Straightforward			
Accountability			
Alignment			
Momentum			
Morale			
Sustainability			
Scalability			
Stylishness			
Learning			
Emphasis			
Balance			
Success Is Inevitable			

When filling out the template, no team should try to populate every cell. The most valuable information is what comes to mind first. A Lessons Learned asset that records three comments is respectable. An asset that records six comments is excellent. An asset that records more than that changes in nature from a tool for proposing action to a place where team members just feel heard.

Skeptics of the template feel its size is overkill. If adopting all fourteen rows at once is too ambitious, choose those within your team's comfort zone instead. Rows of the template are sequenced according to urgency, so the most conservative team might adopt only the first four rows. A more ambitious team might adopt the first seven rows. Long term, any diligent team can handle reflecting on all fourteen culture traits.

Different skeptics dislike Lessons Learned exercises altogether because they feel any negativity is an embarrassing reflection on themselves. Their discomfort can lead to others' discomfort and prevent a team from reflecting, learning, or improving. Predictably, this kind of team has an unsafe culture, and failures are inevitable.

A healthy sign from one Lessons Learned exercise to the next is a shift in what the team identifies. For example, a team might identify concerns about accountability and alignment in July. If, in August, those concerns fade only to be replaced by concerns about momentum and sustainability, the shift shows attentiveness.

Another common habit with a Lessons Learned exercise is to conduct it only at the end of a project. But this lets bad habits survive that much longer, damaging the entire project. Teams should reflect during the project to correct bad habits (and commend positive habits) sooner, not later.

Healthy habits for the Lessons Learned asset assign and execute Five Verbs although in a different manner than for the previous assets. To avoid the buildup of past content and to reduce bias from past remarks, assigned team members and invited stakeholders start with an empty template every month and *draft* a new asset. Each person can review past assets to see whether past comments remain important enough to repeat in the current month's exercise. If a second meeting is warranted, assigned team members *review* and *revise* the asset for the month at that time.

The team might disagree on the exact language that gets memorialized in the Lessons Learned asset. When this happens, the person with the assignment *approve* serves as a tiebreaker to decide on the written content for that month before *distributing* the asset to relevant noncontributing stakeholders. The most commonly impacted downstream asset is the Project Plan (specifically, assignments within the Project Plan).

The Lessons Learned asset is an excellent example of "culture disguised as a template." Monthly, the asset repeatedly shapes your culture's discipline. Conclusions that increase work improve *quality*. Conclusions that decrease work reduce *waste*. The exercise reduces *autonomy* and shows that successful teams have high interdependence. The asset also shapes your culture's empathy, resembling monthly *rehearsals* that reinforce *trust*, *harmony*, and *stewardship*.

Approachability Menu

Coaching individuls in three communication
channels (meetings, email, and assets)

Healthy teamwork requires frequent feedback so that every team member continually contributes what's best for the team. Teams need a simple tool to administer this feedback. The Approachability Menu sets the table for team members to be approachable to each other and coached with as little documentation as a checkmark on a menu.

Conventional performance feedback for individuals often has much better *intent* than *effect* and *impact*. The feedback might reflect personality incompatibilities driven by egos, insecurities, or biases of the feedback giver or recipient. Feedback within a typical team is at the mercy of the team's methodology, which likely suffers from VUCA and perpetual communication traffic jams. The combination of this Methodology Debt and ineffective feedback routines sets individuals up to fail. The Approachability Menu's scope, structure, and frequency encourage pragmatic and mild feedback. The asset encourages team members to share feedback before the behavior becomes a big problem and the feedback becomes emotionally charged.

The Approachability Menu sets the table for individuals to circulate feedback for the three communication channels—meetings, email, and assets. Feedback about meeting etiquette might involve punctuality, interruptions, or monopolizing. Feedback about email etiquette might involve responsiveness, length, or audience fit. Feedback about assets involves assignments of the Five Verbs, duration of work, or a bigger issue with the Project Plan. Customize the menu to best fit your team and try to keep the menu stable so colleagues know what behaviors matter and don't feel surprised.

The menu prompts three potential responses regarding specific instances of behavior: more, less, and an optional explanation of what the person observed, their reaction, and their suggestion. The following example asks for fewer interruptions in meetings, more responsiveness to email, and assigning a person to approve an asset instead of a recent assignment to draft.

Area of Attention	More	Less	Optional ORS (Observation, Reaction, Suggestion)
Meetings			
Punctuality			
Interruptions		✓	I observed Tracy interrupt Robin four times. This disrespects Robin; I suggest that Tracy commit to letting people finish.
Participation			
Email			
Responsiveness	✓		Some important emails to Tracy don't get a response until almost a week later. I wonder if Tracy is buried, and her workload is too high to keep quality high.
Length			
Audience fit			
Assets			
Draft		✓	By drafting training materials, Tracy reduced involvement from her team. I suggest that Tracy delegate this to a junior employee.
Review and Revise			
Approve	✓		I did not observe formal approval of the training materials, so I'm concerned about accountability. I suggest that Tracy formally approves the training materials.
Distribute			

Some skeptics of the Approachability Menu embrace traditional, formal annual performance evaluations led by their Human Resources department. But many employees resent these formalities, and the processes are high maintenance and rife with surprises. In contrast, completing the Approachability Menu is not time-consuming. The conversations are usually short and frequent enough to keep the gravity light. The stability and simplicity of the asset as well as the frequency of conversations all keeps surprises low.

Other skeptics feel frequent coaching is intrusive and no different than micromanaging. But the Approachability Menu is *micromentoring*. It provides unambiguous feedback on a narrow (but customizable) set of teamwork behaviors.[13] Behavior changes resulting from the Approachability Menu are not outrageous sacrifices. Small improvements in one person's behavior can translate to large performance improvements for a team. The menu is a modest but effective tool to ensure all team members are approachable

Like the Lessons Learned asset, the ideal frequency for an Approachability Menu asset is monthly. Its primary downstream asset is the Project Plan—specifically assignments within the Project Plan.

Healthy habits for the Approachability Menu assign and execute Five Verbs. An employee's manager assigns team members to *draft* from an empty template to avoid the buildup of past content and reduce bias from past remarks. The manager *reviews, revises,* and *approves* the asset once they are satisfied with the contributions. The manager *distributes* and discusses the asset with each employee. Employees agree with their manager about appropriate behavior changes and talk directly to the contributors. These conversations continually improve *approachability* in the team's culture.

A blank Approachability Menu has meaning, too; that is, the contributors feel that the recipient's meeting etiquette, email etiquette, and asset contributions are relevant and healthy. Written feedback is not about personality

---

13 More ideas for meeting etiquette include listening, staying on topic, multi-tasking, availability, initiation, RSVP responsiveness, facilitation, and focus. More ideas for email etiquette include initiation, asset fit, clarity, tone, audience fit, spelling, passive voice, fit-for-meeting (instead of email), and fit-for-asset (instead of email).

and the tool aims to neutralize personality conflict. The feedback relates to behavior and sets the table for the recipient to explain their behavior. Differing points of view about behavior qualify as task conflict.

When someone has feedback that doesn't fit into the Approachability Menu, that feedback is usually late, emotionally charged, and a big deal. Go back in time and identify the first signs of the problem. What's likely is that a team member suppressed their feedback when it was still pragmatic and mild. It is never too late to return to the original perceptions and concerns and coach on them because that initial feedback fits in the Approachability Menu, neutralizes personality conflict, and emphasizes behavior and task conflict.

The Approachability Menu is another excellent example of "culture disguised as a template." The asset repeatedly shapes your culture every month. It shapes your culture's discipline by reducing *wasteful* feedback and improving the *quality* and *ease* of meetings and email. The asset shapes your culture's empathy as a *micromentoring* tool so that every employee *has each other's back*. The menu improves *self-awareness* about communication etiquette to *bring out everyone's best*.

Lens of the Individual

Where there is a free press, the governors must live in constant awe of the opinions of the governed.

~ Thomas Macaulay, 1ˢᵗ Baron Macaulay (1800–1859),
British historian and politician

E very employee contributes to the team's work. Every employee also has opinions on the team's health and ideas for improving it. A handful of assets capture this information through the lens of the individual: Individual Status Report, Stoplight Report, I Like I Wish I Hope I Wonder, Workload Report, ORS Report, and Pie Chart.

A typical team culture doesn't ask every employee to be as thoughtful as possible. Average individual contributors focus on doing the work, not judging the work's health. But that doesn't stop employees from forming opinions. Shrugging at thoughtfulness and points of view prevents the rest of the team from benefiting from constructive ideas and mentoring counterproductive ideas. Often, the best ideas about work come from the most thoughtful people doing the work.

Every employee maximizes their value when they organize and prioritize their work and stay attentive to their assignments and new information. Lens of the individual assets accomplish that by capturing recent work, near-future work, health, and sustainability of an employee's work. The assets encourage every manager to listen to their employees and then respond and act to keep employees' value high.

Skeptics might feel that these assets equate to micromanagement. Instead of transparency, doubters prefer freedom, privacy, and ambiguity about their work. This hurts teamwork. Neither do skeptics want to hear what junior employees have to say. Skeptics make employees feel unsafe to share concerns. But these assets demand that leaders listen. These assets demand that leaders allow followers to feel safe to share ideas and concerns without retaliation.

The transparency and safety in these assets build accountability, self-confidence, and courage. They reinforce to employees and managers how both groups can maximize the value of their relationship, collaboration, and teamwork. Differences of opinion between an employee and their manager are valuable because defending opinions improves information sharing and reduces blind spots, which optimizes work in the bigger picture. These assets manage the expectations and perceptions of both followers and leaders.

The following sections sequence the assets according to how common they are in innovation teams (with the most common described first). The Five Verbs framework is applied differently for these lens of the individual assets: A junior employee *drafts* them. Their manager *reviews* them. *Revisions* are rare, and formal *approvals* are counterproductive. If the manager has feedback or questions on any asset, it's an excellent excuse for a meeting to share more complex information than what the asset itself contains.

The default frequency for these assets is weekly. Every few days an innovation professional has important new information to share, but formalizing the new information daily is laborious, and formalizing the new information monthly allows problems to survive for weeks.

These assets shape a culture of discipline in a few ways. By exposing *waste*, these assets help minimize it. The assets manage *autonomy* by showing the range and boundaries of what every employee works on. By managing the *idleness* of work or workers, the assets maintain *elasticity* for a team.

These assets shape a culture of empathy by encouraging *co-creation* and reinforcing *trust* between employees and managers.

Individual Status Report

> A team member's report on primary accomplishments from the past week and goals for the upcoming week.

An Individual Status Report is the most fundamental asset to capture the lens of the individual. In a steady rhythm, every employee should generate a formal status report for their manager that shares the information their manager should know. The report helps a manager gauge the health of their employee's work and raise its long-term value.

A typical environment handles status informally. Agile environments require employees to conduct short daily meetings with their team. Traditional environments suggest that employees have a one-on-one conversation with their manager approximately every week. These habits are respectable, but an informal chat alone doesn't maximize discipline, empathy, and value. A status report does.

A status report encourages employees to describe the information succinctly and precisely. An innovation professional needs ten to twenty minutes to craft a status report, and a manager needs three to five minutes to read it. These two steps share more information than an informal status conversation and don't require synchronizing schedules. A carefully worded status report distinguishes simple from complex information so that any status conversations can skip low-ambiguity details and jump to ambiguous and complex topics. Because of this, status *reports* make status *meetings* more valuable.

A generic example of a status report is located on the next page.

Workstream	Accomplishments and Highlights of Past Week	Plans and Goals for Next Week
Task A	Completed W	Start Z
Monthly Process B	Continued X	Complete X
Project C	Started Y	Continue Y

On the report, an employee lists the major workstreams in the first column. In the second column, they list their accomplishments since the last status report. On a slow week without a legitimate accomplishment for a certain workstream, the employee shares any kind of "highlight." In the third column, they list plans and goals for the next week. The employee specifies the assets to which they're contributing and which of the Five Verbs (draft, review, revise, approve, and distribute) characterizes their asset contributions. They minimize ambiguity by using words like *start*, *complete*, and *continue*. For example, a report might state for last week, "Completed Draft of Training Approach" or "Continued Reviewing and Revising Elevator Pitch." For *next* week, the report might say, "Start Approving Test Script" or "Complete Reviewing and Revising Migration Script."

Another long-term benefit of status reports is their value to an annual performance assessment. Months of status reports help employees and their managers recall and summarize an employee's contributions for the year. Status reports are a form of receipt—proof that something happened in the way it was recorded weeks or months ago.

Skeptics of a written status report are not interested in memorializing accomplishments; in fact, they are interested in *not* memorializing them. At best, such a skeptic is embarrassed about the low value of their work. At worst, the skeptic's work is unethical and damaging to the company. Skeptics' *values* undermine employees' *value* to the organization. Skeptics are not interested in maximizing discipline or empathy between employees and their managers. To the skeptics' dismay, Individual Status Reports minimize misunderstandings and maximize common understanding about priorities and progress. Individual Status Reports govern a virtuous circle of self-esteem, autonomy, and transparency.

Managers are busy people. One-on-one status meetings are the easiest events to cancel, so canceling is common. But neglecting sharing status reliably leads to negative surprises. Status reports ease managers' lives because managers learn the information in a fraction of the time of a status meeting and can respond thoughtfully. The report signals to a manager how much or how little attention they should give to the employee's work to balance their speed, quality, and autonomy.

At the same time, the asset provides employees with a sense of accomplishment and boosts self-confidence. A status report helps employees "manage up" without "kissing up." The report is also a professional way to escalate a problem and ask for help.[14]

The Project Plans containing employee assignments are the parent innovation assets of an Individual Status Report. Parent operational assets of an Individual Status Report are Current State Process Flows, which show employee assignments. Children assets of the Individual Status Report include the other lens of the individual assets: Stoplight Report, I Like I Wish I Hope I Wonder, Workload Report, ORS Report, and Pie Chart.

A healthy frequency for an Individual Status Report is weekly. An employee *drafts* the report and *distributes* it to their manager. It's possible that the manager's *review* of the report generates feedback or questions and justifies *revising* a status report. A mutual understanding of the report's content is a proxy for *approving* it. Diligent employees store their Individual Status Reports throughout the year to jog their memory for their annual performance review. When team interdependencies are complex or misunderstandings could be crippling, employees might *distribute* their status reports to their team members.

The Individual Status Report asset shapes your culture's discipline because it motivates every employee to maximize the *quality* of their work every week.

---

14 One rule of thumb is to escalate any decision or situation where there is more than a 10 percent chance your manager disagrees or disapproves.

The report *eases* a manager's job. The report shapes your culture's empathy because it encourages every employee to show they are a *good follower*, *self-managing*, and *self-sufficient* in reporting their contributions to the team.

Stoplight Report

Crude health assessment to determine whether deeper troubleshooting is worthwhile.

A Stoplight Report provides simplistic "health at a glance." It doesn't exist independently; it piggybacks five assets: Scorecard, Lessons Learned, Workstream Status Report, Individual Status Report, and the Use Case Assessment. To convey health rating of work, assign the colors of a stoplight—Red, Yellow, and Green (RYG)—to ingredients of other assets.

Stoplight Reports are common in businesses and innovation teams but are often used too lightly or too heavy-handedly. Using a report too lightly is a team rating something Red or Yellow, but not thoughtfully identifying, planning, and executing a path to return it to a Green rating. If something in your Scorecard or Use Case Assessment is Red or Yellow, the path to Green involves the Change Log, a Project Charter, and revising a process flow. If something on a Lessons Learned exercise or status report is Red or Yellow, the path to Green likely involves changing an assignment in the Project Plan. In the Stoplight Report, include your next steps for returning to Green ratings.

Heavy-handed use of the Stoplight Report applies the objective standards for each color religiously, prohibiting exceptions and suppressing a rationale for an exception. Rare occasions can exist where nuances matter more than simplicity and where subjectivity should rule over objectivity. Teams disagreeing on color is common, making the role of tiebreaker important so the team can move on. Debating health-at-a-glance is less important than converging on a change to the Project Plan that addresses a concern. It's common for parties who disagree about Red, Yellow, and Green to agree on a proposal from the pessimistic party.

Assigned team members regularly disagree about some aspect of project health (color) as they *review* and *revise* a Stoplight Report. The Stoplight Report is the perfect place to exercise the power of diversity in a team. Physical and financial health matters often benefit from second opinions. The Stoplight Report is a simple, straightforward, pragmatic tool to leverage a team's diverse expertise and risk profile to uncover blind spots and risks. The Stoplight Report fosters *disagreement* but undermines *demonization*. It welcomes *task conflict* and neutralizes *personality conflict*. The health-at-a-glance nature of the report is a baby step for identifying the proper response to bring the team closer to consensus Green ratings. A Stoplight Report draws perspectives from *individuals* but pertains to *team* health, so it steals from competitive energy and feeds collaborative energy.

The next page shows an example of a Stoplight Report applied to a Lessons Learned exercise for two projects.

When a team populates a Lessons Learned asset, it also assesses its health as Red, Yellow, and Green. A team should populate the Stoplight Report *only enough to uncover one or two Red or Yellow traits*, just enough for a team to work to return to Green by the next month. Revealing a half-dozen Red traits can overwhelm a team. The report contains a placeholder to explain a color change and the plan to return to a Green rating. If any dimension of the project is Red or Yellow, the team should discuss that dimension every month until the team considers that dimension Green again. Returning a trait to Green encourages a team to populate more of the report, which can uncover additional Red or Yellow culture traits so the team can repeat the process to return to Green for additional culture traits.

Skeptics of the Stoplight Report misuse or abuse it by never changing colors. Some *misuse* the report when they consistently report everything as Green, which is typically a sign of low rigor, low humility, and high ego. Some *abuse* the report by mindlessly keeping items Red or Yellow for weeks without proposing a solution to return to a Green rating. No desire to return rating to Green signals a lack of ability and accountability, whereas colors changing reasonably often indicates that the team is attentive, thoughtful, and humble.

Area	Color for Project A	Color for Project B	Placeholder for Plans and Explanations
Safety, Inclusivity, Belonging	Green	Red	Revise assignments in the Project Plan
Transparency	Green		
Simple and Straightforward	Red		Revise Future State Scripts and Future State Process Flows
Accountability	Green	Green	
Alignment		Yellow	
Morale		Yellow	
Momentum	Yellow		
Sustainability			
Scalability			
Stylish			
Learning			
Emphasis			
Balance			
Success feels inevitable			
Placeholder for Plans and Explanations	Placeholder	Placeholder	

Instead of viewing them as upstream or downstream assets, four assets qualify as parent assets: Scorecard, Lessons Learned, Workstream Status Report, and Individual Status Report. The frequency of a Stoplight Report should match that of its parent asset.

A Stoplight Report follows the Five Verbs of its parent asset; when all contributors agree on colors, the report cruises smoothly through *draft, review, revise, approve,* and *distribute.* When a team does not take agreement for granted (and it shouldn't), agreement on colors in the Stoplight Report is reassuring; the accountability in alignment makes a strong public statement. Modest levels of disagreement or diverging points of view (i.e., good-faith disagreements, not abusive ones like the example above) show attentiveness, quality, and care among team members. Consistent consensus and thoughtless *agreement* on Red, Yellow, and Green can be signs of low diversity and lack of attention and care. Whether or not members agree on Red or Yellow ratings, the team *drafts, reviews*, and *revises* the next steps to return to Green rating. The team member assigned to *approve* often plays tiebreaker when the rest of the team does not reach a consensus. The team *distributes* the parent asset to interested noncontributing stakeholders.

A Stoplight Report shapes discipline by prompting teams to *vigilantly* explore ways to improve the *quality* of the work and the employee experience. The report shapes empathy by asking teams to *listen* to individuals' worries and leverage *diversity* in a team's skills and perspectives. It fosters *humility* that empowers team members to rate work Red or Yellow and requires *resilience* for them to return work to Green rating.

I Like I Wish I Hope I Wonder

Captures positive sentiments, negative sentiments, questions from the curious, and ideas from the dreamers.

A person finishing any of these four sentences tells you a lot about what they think. They could be sharing compliments, ideas to improve team health, or concerns with the assets. People's responses expose perceptions and expectations that need managing. Diligent employees append this information to their Individual Status Reports.

A typical manager explores the minds and opinions of their employees too little because of presumptions of intellectual superiority, power preservation, or fear of what they might hear and learn. Underexploring hurts trust and alignment, as well as the quality of information sharing. Good managers respect, trust, and listen to their employees' perspectives. Good managers fear what they *don't* learn more than what they *do* learn from their employees.

Employees don't have to *finish* these sentences every week, but thoughtful employees at least *reflect* on them weekly. Procrastinating on making these opinions known reduces their value and perpetuates unalignment and misperceptions for weeks. Proactive reporting moderates the quantity and strength of opinions.

The four sentences try to reveal positive and negative impressions and the neutral reflections of human minds. The sentences try to keep employees concise. Finishing the sentence "I like" means the employee wants to share a positive reaction to an incident, accomplishment, or culture trait. Finishing the sentence "I wish" means the employee prefers that something in the past was different. Finishing the sentence "I hope" means the employee prefers that something in the future proceeds in a certain way. Finishing the sentence "I wonder" means the employee wants to pose an idea without restraints or a preference for or against it. This could be an innocent question or a politically sensitive topic.

Examples include:

"I like the enthusiasm I saw among the testers."
"I like the thoroughness and tough decisions I see in the Project Charter."
"I wish the webpage wasn't so busy."
"I wish this code didn't feel throwaway given what's next on the Roadmap."
"I hope our test data is ready in time for the start of test execution."
"I hope training sessions will include all offices."
"I wonder if training sessions will be in person, online only, or hybrid."
"I wonder if we have to purge data related to the divestiture."

A manager has three options upon acquiring their employees' sentences. In the first scenario, the manager agrees with the employee's view and will do something about it (coaching a team member or scrutinizing the relevant asset). In the second scenario, the manager has the same opinion but feels it does not warrant action. In the third scenario, the manager explains their contrasting view to the employee and defends it. In every case, good managers outwardly respond to these "opinions of the governed."

Skeptics of the I Like I Wish I Hope I Wonder report prefer to suppress their employees' attentiveness and ideas. This reduces employees' value. It also creates a toxic environment and a culture of low safety. Good managers, in contrast, enjoy leveraging their employees' attentiveness and ideas and want a high-safety culture. Good managers know that ignoring their employees' opinion sharing means the opinions go elsewhere—neglected or managed by others.

The parent assets of the I Like I Wish I Hope I Wonder report are the Individual Status Report and the Lessons Learned exercise. The most common downstream assets of the report are the Project Plan and the Parking Lot. Instead of separate execution of Five Verbs for team collaboration, every employee includes an I Like I Wish I Hope I Wonder report with their Individual Status Report so they can share this information with their manager weekly. Teams apply I Like I Wish I Hope I Wonder within a monthly Lessons Learned asset.

The I Like I Wish I Hope I Wonder report shapes discipline by *vigilantly* managing employees' *autonomy* of opinions. The report shapes empathy by creating a *safe* and *authentic* communication tool to be shared between employees and managers. When a manager wants to defend against an employee's opinions, the report helps the manager *immobilize* and *reject* counterproductive ideas that should not proceed while encouraging the *diversity* of ideas from all employees.

Workload Report

A report that captures a sense of workload, overload, or ability to take on more work.

A Workload Report tells a manager whether an employee feels the volume of their assigned work is appropriate and sustainable. Managers' subsequent balancing of workloads across many employees keeps entire workstreams valuable and sustainable.

A typical environment doesn't explicitly manage workloads. Informally, employees might tell their manager their workload is too high or too low. If they do, it hints that their workload is *egregiously* too high or too low. Or a manager might notice unexpected behavior or a change of pace or quality of work and wonder whether an employee's workload is too high or too low. But in a typical environment, managers underreact, and employees persevere with the status quo. Opportunities to increase an employee's value (when their workload is too low) or relieve a bottleneck (when their workload is too high) are missed, and managers often overlook workload extremes among employees.

A Workload Report requires about ten seconds to piggyback onto a weekly Individual Status Report. An employee rates their workload on a five-point scale.

5 = Workload is too high and not sustainable | I'm a bottleneck for a workstream | Idle work exists | Recommend reducing my workload
4 = Workload is full | Increasing workload might create a bottleneck
3 = Workload feels healthy and sustainable | Short burst of additional work is OK
2 = Workload can handle additional new workstream | Idle worker
1 = Workload is very low | Idle worker | I feel excluded and under-utilized | I am concerned about my job security

The time-consuming work that extracts value from a Workload Report sits with managers. Managers determine what to do with the Workload Reports of employees who report to them. Good managers want to optimize the long-term value of workstreams. To do this, they play matchmaker between workstreams and workers.

Across a team, it's common for someone's workload to become too high or too low. Having these extremes is not a crime (perhaps a sampling error). *Keeping* the extremes as systemic or systematic errors—suppressing or ignoring information that eases bottlenecks and improves sustainability—is a crime. Good managers keep themselves informed of workload extremes and use the information to keep work moving and employees valuable. To accomplish this, good managers routinely rebalance workloads.

Approach rebalancing across two dimensions. First, determine whether the imbalance resides in operations workstreams or innovation workstreams. Second, determine whether the imbalance is local or global. Local imbalances resemble sampling errors that require bottom-up solutions. Global imbalances resemble systemic or systematic errors that require top-down solutions. The following table shows the four permutations of imbalanced workloads and the respective primary downstream asset to *revise* to restore balanced workloads. Hiring and releasing workers is one lever for changing workloads, but it's best as a secondary response to workloads that are too high or too low.

Four Impacted Downstream Assets		Workstream Type	
		Operations	Innovation
Altitude of Imbalance	Global	Customer Experience Hierarchy	Roadmap
	Local	Current State Process Flow	Project Plan

For example, if the reports show that innovation workloads are too low globally, revise your Roadmap, find and start new projects, or release some workers. If reports show innovation workloads are too high locally, reassign the work in the Project Plan away from the bottleneck employee to available workers.

Skeptics don't want to learn, understand, expose, or address imbalances related to workload or personnel. They are comfortable allowing workloads to remain too high or too low—with high idle work and idle workers. Neglecting to matchmake and tolerating unsustainable workloads equate to terrible resourcefulness and management. Matchmaking and workload balancing comprise excellent resourcefulness and management.

Other skeptics worry that employees might be dishonest and overestimate their workload. If an employee's status report looks sparse, an overloaded employee is being either unproductive or dishonest. But if an employee's status report looks full, an overloaded employee is likely being honest. Honest or not, when an employee reports their workload is too high, reassign work away from them toward an available employee.

Instead of an upstream asset, it's more accurate to characterize the Individual Status Report as the lone parent asset of the Workload Report. As explained above, the Workload Report can have up to four downstream assets: Customer Experience Hierarchy, Roadmap, Project Plan, and Current State Process Flow. Instead of separate execution of Five Verbs, every employee includes the Workload Report with their Individual Status Report and shares this information with their manager weekly.

The Workload Report (and subsequent matchmaking) shapes your culture's discipline. The report keeps levels of *idle work, idle workers,* and *waste* low. This improves *speed* and *elasticity*. The report shapes your culture's empathy by keeping workloads *inclusive* and *sustainable*, *balancing* responsibility across many employees. It reveals ways a manager can be a good *steward* of everyone's time and talent.

ORS Report

Framework to share spot feedback in the form of observation, reaction, suggestion.

Occasionally, employees have feedback that is unsuitable for the sizable

Lessons Learned and Approachability Menu exercises. Feedback doesn't always fit inside a weekly or monthly routine. The ORS (observation, reaction, suggestion) framework structures such information to include a constructive idea with enough context to avoid passing along simplistic criticism.

Ambiguous feedback has low value to its recipients. Examples of vague feedback include "That didn't work for me," "They're difficult to deal with," and "They're not a team player." The words might contain plenty of truth, and uttering them might make the speaker feel better, but thoughtless grievances are of minimal value to anyone hearing them. Systemic and systematic ambiguous feedback (thoughtless grievances) reflects on the speaker, not the recipients. Governing feedback with the ORS framework discourages thoughtlessness and raises the quality of feedback.

Another scenario for the ORS Report is when diverse feedback is most valuable. Most innovation teamwork focuses on collaboration toward converging ideas, and the ORS framework is uniquely valuable for centralized decision-making that welcomes diverging ideas. The report accepts conflicting and contradictory ideas and leaves it to one person to implement what they see as the best combination of suggestions.

The ORS Report also encourages the feedback giver to be concise. Below is an example.

Observation	I noticed that people in the meeting looked nervous and were looking down at your document while you were speaking.
Reaction	A lot of us, including me, felt unprepared.
Suggestion	Either circulate the draft 1–2 days before the meeting, or reserve 10 minutes at the start of the meeting for us to read in silence, or both.

Skeptics of the ORS framework dislike its clarity. They prefer to give feedback that's easy for them to share. But the framework reinforces the idea that everyone has the opportunity and responsibility to be attentive to their colleagues and to help them improve. Riffing on a common "train the trainer" approach, the framework is a "coach the coach" approach.[15]

The ORS Report lowers the barrier to sharing ideas. Skilled and well-intentioned employees might suppress feedback if they feel their idea is petty. Because the framework structures feedback to be concise regardless of the significance of the suggestion, employees are more comfortable using it to share simple or superficial feedback points. As we know, small changes to written or spoken messages can significantly impact audiences and stakeholders. The framework is especially effective in group settings because hearing major or minor feedback warms up an audience to generate more feedback.[16]

Feedback recipients are likely to capture highlights in writing, but unlike other project-independent assets, the ORS Report doesn't emphasize documentation, alignment, frequency, or sequence. Neither does the ORS framework fit with executing Five Verbs to draft, review, revise, approve, and distribute feedback because, especially in group settings, feedback recipients typically receive diverging, conflicting, and contradicting ideas from their colleagues.

The ORS framework shapes your culture's discipline by improving the *quality* of the speaker's feedback and the recipient's work. It reduces *wasteful* feedback. The framework shapes your culture's empathy because team members' concise, thoughtful feedback "brings a brick, not a cathedral" (of feedback) and conveys to colleagues "I got your back."

---

15 A train-the-trainer model builds a two-step process to inexpensively disseminate information to dozens of people while building the skills of the trainers and allowing them to tailor their message.

16 If a message is less about feedback and more about persuasion, consider the three-point framework of "What? So what? Now what?" Tell the audience the important topic, explain why they should care, and prescribe the next steps for them to contribute to a solution.

Pie Chart

A crude metric to reduce time and cost of
firefighting and undesirable operations.

Most innovation professionals spend less than 100 percent of their time on project work. They spend time on routine (operations) and unplanned (firefighting) work. Managers should know how their employees spend their time, or *perceive* spending it, across innovation, operations, and firefighting. Periodically, a manager compares and aligns with the employee on this time distribution to put healthy pressure on the work distribution. Over a few months, changes in the Pie Chart reflect changes in the employee's long-term value and morale.

Innovation work is *expected* work. It builds a new process or capability that disrupts the status quo. Project Charters and Project Plans govern innovation work. Operations work is also expected work, but it is the status quo. It has a rhythm and often a daily, weekly, or monthly frequency. Current State Process Flows govern operations work, and a Scorecard reports time, duration, or quantities of operational transactions. Firefighting, however, is *unexpected* work, work without rhythm or steady frequency. No assets govern firefighting, but firefighting is valuable and necessary when it helps avoid a more significant crisis.

A typical organization doesn't track or manage this information. It's difficult to quantify firefighting information top down or know what to do with macrolevel information. It's demeaning for a single person to quantify this information bottom up (precision is unimportant) then explore and diagnose microlevel changes to improve the distribution of the work. The Pie Chart is a decentralized approach to monitor the work distribution and identify changes that will raise the value of work.

The worst-case scenario is that an employee spends a high percentage of time firefighting and a low percentage of time on innovation. This is the worst case because firefighting is high-cost work unable to be governed by a process flow. Every cost-minded manager wants to find the origin of their employees' firefighting and implement processes that nullify it (and, yes, this might require innovation, i.e., a project). Over a few weeks or months, executing firefighting prevention processes shifts work away from firefighting and into operations.

When an employee's firefighting level is satisfactorily low, their percentage of time spent on operations work is likely high. A cost-minded manager inspects their employees' status reports for the operations to which they contribute. The manager reviews the Use Case Assessment (detailed in the next section, "Current State") to see which processes can benefit most from innovation. A match between an employee's time-consuming operations work and a problematic use case is a great candidate for innovation. The employee and manager should then add the candidate to the Change Log.

The short-term goal of the Pie Chart is to minimize firefighting for every employee. The long-term goal of the Pie Chart is to reduce operational costs ("streamline operations") to a level fitting the company mission, customer intimacy, and innovation intensity. A cost-minded manager is constantly heedful of reducing costs related to firefighting and operations.

When first adopting the Pie Chart, it's common for an employee and a manager to disagree temporarily on whether certain work is innovation, operations, or firefighting. If the employee's work appears on a Project Plan, it's innovation. If the employee's work appears in Current State Process Flows, it's operations. Otherwise, the work is firefighting, and the manager and the employee should document the current state process.

Figure 7 shows pie charts and transitions between the short-term and long-term goals. Differences between Charts 1 and 2 show an employee who reduced their firefighting from 40 percent of their time to 10 percent. Differences between Charts 2 and 3 show an employee who reduced their operations workload from 80 percent to 40 percent. Precision in the charts is not important. What's important is that the charts influence behaviors and show a healthy trend of reducing firefighting and operational costs.

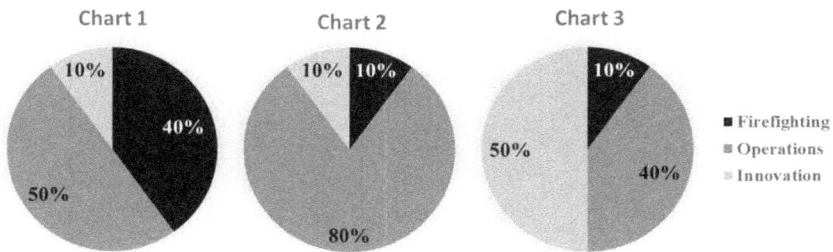

Figure 7. Pie chart.

Skeptics don't want to expose high-maintenance operations, chaos, or mismanagement related to firefighting. Some employees and managers invite and *enjoy* chaos—even to the point of it being a personal value. Skeptics need to realize that the Pie Chart asset doesn't reduce work. It *shifts* work—first from firefighting to operations, then from operations to innovation. Innovation is a form of exploration that involves uncertainty and hard work. Skeptics should convert their affinity for firefighting and hard work into a passion for the wealth of customers' innovation needs, their sense of urgency, and the profits that accompany it. The Pie Chart is a simple tool for every employee to reduce costs, increase innovation, and improve profits.

The Individual Status Report is the Pie Chart's lone parent asset. When an employee does too much firefighting, the downstream asset is Current State Scripts. When an employee's work is primarily operations, the downstream asset is the Change Log.

Instead of separate execution of Five Verbs for team collaboration, every employee includes the Pie Chart with their Individual Status Report in order to share this information with their manager weekly. Disagreements about the three pie chart estimates might expose reasons to change assignments or communication channels but, then again, disagreements might not change anything.

The Pie Chart shapes your culture's discipline by reducing *waste* in firefighting and *variability* in chaos. Reducing operational costs improves a company's *economics*. This asset shapes your culture's empathy by keeping employees' work under *control*, with less *messiness*. The Pie Chart enhances an employee's experience with—and a *story* about—their employer.

Current State

The best time to plant a tree was twenty years ago.
The second-best time is now.

~ Chinese proverb

Before pursuing any project, an innovation team sets itself up for success by gaining clarity and by knowing the details about the current state of the business.

Many large businesses document some flavor of current state information. Common labels include SOP (standard operating procedure), KTLO (keeping the lights on), and BAU (business as usual). Companies that neglect current state documentation build up tribal knowledge that resembles documentation debt. Tribal knowledge slows information sharing and increases risk when employees leave the company.

Current state documentation (CSD) lowers the cost of several valuable innovation tasks, including isolating innovation candidates, partitioning projects into manageable sizes, and prioritizing them. Without documentation, teams need much more talking time to accomplish this work. And conversations about the current state are often intimidating because employees have the incentive to tell others how complicated their jobs are. CSD educates others about intricate detail in a less intimidating way. CSD steers criticism toward processes and away from people, jobs, and decisions. It reduces inter-rogative and agitating conversations and fosters unhurried thoughtfulness. Seeing process imperfections in CSD inspires passion and purpose for innovation and builds confidence that the team can succeed.

Skeptics of CSD are threatened, humbled, or even humiliated by the transparency of it. CSD exposes the inevitable flaws in bad and neglected processes. It's rational that skeptics want to avoid looking bad, and low transparency allows them to keep their tribal knowledge, job security, and personal competitive advantage within their team. But low transparency is a competitive *disadvantage* for the team. Building and viewing CSD can be painful, but as Laurence Peter, a professor of education at University of Southern California, remarked, "Nothing will change until the status quo is more painful than the transition." The best time to document CSD was during the project that could have called it *future* state documentation. The second-best time is now.

Other skeptics worry about suspending all innovation work to catch up on CSD in a single burst of activity. Just as it's painful and unrealistic to pay back financial debt with a balloon payment, it's unnecessary to pay back documentation debt at some unnatural pace. It's sufficient and most sensible to pay back documentation debt steadily, in a sequenced manner that keeps ahead of what projects are likely next.

CSD is not glamorous work when done as remedial, post-project paying off of documentation debt. A silver lining is that CSD is an opportunity to make new employees unmistakably valuable. A perfect—and the easiest—task for a new employee is executing Five Verbs for CSD. A new employee *drafts* and experienced employees *review*. The new employee *revises*, and experienced employees *approve*. The new employee then *distributes* the CSD to the relevant stakeholders.

Other than for the Scorecard asset, the question of frequency for CSD doesn't apply. A team creates it only once, and that is when it is in docu-mentation debt.

Project-independent CSD shapes your culture's discipline by raising the *autonomy* of individual projects and improving the *speed* and *ease* with which their assets are created. CSD shapes your culture's empathy through *authenticity*, *messiness*, and *humility* in the short term and *self-confidence* in the long term.

Scorecard

> A list of what an organization measures about its operations, recent historical values, and target values.

Stakeholder value and operational health can often be measured. The purpose of measurement is to generate ideas for improvement and innovation, and measuring a few operational traits is enough to accomplish this. The asset to record measurements is a Scorecard.

A typical *operations* team records metrics, but many *innovation* teams do not use metrics as a primary input to innovation ideas. A Scorecard is a straightforward tool to infuse pragmatism into proposing innovation ideas. Metrics typically aim for objectivity, although some subjectivity accentuates the company's mission and values. A healthy team is aware of—and honest about—subjectivity and biases, and a Scorecard transparently presents what an organization sees as valuable.

Even if data is accessible and plentiful, consider starting small. Maintaining a scorecard is not free, and overdoing it undermines morale. A rule of thumb for an appropriately sized Scorecard is to include only enough items to generate a few innovation ideas (which translate to new items on the Change Log). Consider using metrics where changing them feels low effort and high impact. When adding a Stoplight Report to your Scorecard, consider metrics with a Red or Yellow rating. Manage metrics that reflect poor operational health or low stakeholder value until you complete the innovations, or when the new processes move metrics, or when the stoplights return to Green ratings. Once a metric feels firmly Green, replace it on the Scorecard with a Red or Yellow metric—constantly measuring for the best innovation ideas. Metrics that matter "move the needle."

A Scorecard requires vulnerability and humility at first. Months of rigorously managing it improve self-confidence and morale. When your Scorecard and metrics-gathering processes are mature, a larger Scorecard is easier to maintain. A list of metrics with Green ratings indicates high stakeholder value and good operational health.

Metrics have two categories—leading and lagging—that describe how much you control—manipulate or monitor—them. Differences between leading and lagging metrics often sit along a *spectrum* of control; they are not simply one or the other. Leading metrics are what your organization has high control over via a decision and without a process change. Treat leading metrics as levers to push and pull or as dials to rotate left and right.[17] Ideas for leading metrics include:

- Effort per week

- Number of people

- Number of phone calls

- Number of handoffs

- Percentage of scripts executed

- Quantity in flight (of projects, tasks, widgets)

- Quantity of employees experiencing Event X

Lagging metrics are what your organization cannot directly control. Treat lagging metrics as if they are analog or digital gauges protected

---

17 A thermostat shows numbers that are leading metrics. A thermometer shows numbers that are lagging metrics.

behind glass. When some time has passed after you have moved some levers, go read these gauges. Ideas for lagging metrics include:

- Duration of a process

- Customer wait time

- Quantity of Customer Type A experiencing Event X

- Quantity of Customer Type B experiencing Event Y

- Quantity of total customers

- Average transaction value

- Quantity of exceptions or errors

- Effort on rework

- Quantity of customer incidents or inquiries

- Quantity of referrals

- Turnover/shrinkage

There are two views of a Scorecard. The view for Outdoor Adventures, on the next page, describes health and goals.

In this view, include a title, a description, and whether you view the metric as leading or lagging. If valuable to the team, have a baseline (common and recent) value for comparison to the actual current and future values. Include specific ranges for what you consider Red, Yellow, and Green ratings if you have ranges in mind. Otherwise, note the desired direction, that is, whether you want the value to increase or decrease, and describe—even if ambiguous—an improved measurement, that is, an "expected lift."

The second view, on page 91, contains quantities measured at the end of the month (EOM); for example, for the number of open customer inquiries, duration of camping registration, and employee effort for a camping

Metric Title	Description	Leading or Lagging?	Baseline	Current Range	R/Y/G	Desired Direction	Expected Lift
Open Inquiries	Number of customer inquiries not yet responded to or resolved	Lagging	N/A	20–130	Y	Down	Reduce peaks
Registration Duration	Average hours lapsed from registration request to completion	Lagging	72	80–120	R	Down	Return to 72
Registration Effort	Average minutes of labor across employees	Lagging	120	120–150	Y	Down	Decrease to 100

Monthly Metrics (EOM)	Sept.	Oct.	Nov.	Dec.	Jan.	Feb.	Mar.	Apr.	May	Jun.	Jul.	Aug.	Sept.
Open Inquiries	105	95	65	44	22	19	38	55	89	121	130	113	85
Registration Duration (Hours)	78	81	98	115	111	102	97	93	88	86	81	78	74
Registration Effort (Minutes)	120	125	138	148	135	131	125	119	125	132	139	133	118

registration at Outdoor Adventures. This view includes historical values at specific frequencies (daily, weekly, monthly) and could also include future target values (labeled so).

Skeptics of a Scorecard might publish their team's best-looking metrics, but they dislike transparency for low stakeholder value and poor operational health. Avoiding your worst metrics doesn't cure the worst of your operational health issues, but poor transparency of metrics suppresses valuable input for innovation ideas. Poor transparency creates blind spots and reduces confidence in innovation ideas. Abdicating metrics that matter disables the ability to manage the expectations and perceptions of those functions. A Scorecard is vital for managing expectations, managing perceptions, and curing operational health.

Unless a metric warrants a higher frequency, execute the Five Verbs routine monthly. A junior operations employee gathers information to *draft* the Scorecard. Operations team members *review* and *revise* it, an operations manager *approves* it, and the junior employee *distributes* it to assigned noncontributing stakeholders. The Scorecard's primary upstream asset is Report Detail (explained in the "detailed design" section later). Its primary downstream asset is the Change Log.

A Scorecard shapes your culture's discipline by modeling *vigilance* and improving the *quality* of innovation ideas. A Scorecard shapes your culture's empathy by showing *self-awareness* of low stakeholder value, poor operational health, and metrics that matter. A Scorecard feeds *resilience* by identifying innovation ideas and declares a metric *GETMO* when it returns to a Green rating.

Current State Inventory

A list of existing operational workstreams and any documentation describing current operations.

A Current State Inventory (CSI) is a stepping-stone toward transparency for your business. The CSI is not the dozens or hundreds of actual documents but a list of documents that exist, or *should* exist, along with ownership and health information.

A typical company cuts corners in various ways, such as by forgiving timelines, skipping approvals, or by bending the rules on financial matters. Neglecting current state documentation (CSD) about business processes and customer experiences—documentation debt—is a common example of cutting corners. One goal of a CSI is to identify and prioritize any CSD worth creating new, that is, replacing those corners to set up future innovation work for success.

Before populating your CSI, know what the CSI is not. It is not a list of current in-flight projects (you document that in a Roadmap). It is not a Report Inventory, which is data-centric and deserves separate attention as a separate asset. Finally, a CSI does not suggest moving the documentation to a single repository or "single source of truth." Whether CSD resides in a single repository or a dozen of them, the CSI merely records the location of the documentation.

In populating a CSI, give each document a title and a description. Categorize each document—broadly such as process, technology, or something more specific.[18] Processes can range from customer facing to back office. Technology-oriented documentation usually includes design, build, test, and deployment activities. A separate category specifies whether the documentation already exists. It's more natural to first populate the CSI with documentation that already exists than with documentation that should exist.

Note the document's format, for example, Microsoft Office Word, PowerPoint, or Excel, or state that the document does not yet exist. For a process document, note its typical frequency (weekly, haphazard, continual,

---

18 Because legacy methodologies are software-centric, process assets are typically more neglected than technology assets. Document by document, process has a larger impact on innovation costs, so emphasize it over technology documentation.

or unknown). Note the document's customer (i.e., who cares most about its content) and its owner (i.e., a person who changes it more often than others).

Three pieces of information help prioritize any work on documents: an estimate of the room for improvement of the document (high, medium, and low are adequate); for process documents, an estimate of the room for improvement of the process; and, finally, an estimate of the cost of delay in improving either the document or its related process. The following page illustrates a CSI example.

Skeptics of a CSI typically see the work to build it as remedial and unexciting. That's a fair and understandable perception. Working on a CSI equates to paying back debt—documentation debt. The CSI exposes "bad news," and plenty of business leaders dislike bad news. It also reveals a company's disadvantages related to low awareness, understanding, and enthusiasm about its operations. If you want to reduce your company's weaknesses, vulnerabilities, and tribal knowledge, build and maintain a CSI. If you want to increase your company's awareness, understanding, and enthusiasm for its operations, create and maintain a CSI. A robust bird's-eye view is a catalyst for completing unexciting work so your team can proceed to exciting work. What's more remedial and unexciting than getting out of documentation debt is staying in documentation debt.

Assign a junior employee to *draft* the CSI. Assign the documentation owners cited in the CSI to *review* and *revise* it. As the CSI matures, contributors have less to add and change. When the CSI feels stable, with few changes, senior point persons (explained in a later section) and relevant project sponsors should conduct monthly and project-specific *review* and *revise* sessions. Documentation owners should *approve* the asset monthly and within the Closure Report for every project. Assign a junior employee to *distribute* the CSI to interested noncontributing stakeholders.

The first version of the CSI doesn't have an upstream asset because it doesn't wait on any other asset. The Closure Report is its ongoing parent asset because it instructs the team to merge every project's future state documentation into CSD. The CSI's downstream assets are Current State Scripts.

#	Title or Short Detail	Long Detail	Process or Tech	Format	Frequency	Customer(s)	Documentation Owner	Room for Improvement of the Documentation High Med Low	Room for Improvement of the Process High Med Low	Cost of Delay for Improving High Med Low
1										
2										
3										

A CSI shapes your culture's discipline by increasing the *transparency* of business operations and exposing the *variability* and *sprawl* of documentation. A CSI shapes your culture's empathy by requiring *self-awareness* and *vulnerability* to build it. It is the easiest asset to mobilize *beginners*, who can contribute to the asset with high *self-sufficiency*.

System Actor Inventory

> The view and assessment of your technology portfolio that eases innovation decisions.

Your organization has a suite of software that supports the business. Have the System Actor Inventory handy as you build current state documentation (CSD) and make innovation decisions that add and subtract from the inventory.

The typical technology organization possesses this inventory. Every company has "build/buy/rent" decisions, and the list contains a combination of systems the company has built, bought, or rented. Occasionally, senior employees make technology decisions in the interest of speed, that is, siloed decisions that under-consult other parts of the company. Years later, the costs to build/buy/rent/maintain are different. It's common for a company to realize it has redundancy in its sprawling technology portfolio.[19] Because of redundancies or gaps, the company might make subsequent technology decisions that are less siloed. A System Actor Inventory facilitates globally minded technology decisions. Although the long-term goal of the inventory is to aid in decision-making about replacing systems, the short-term goal is to help build the rest of the current state assets.

For every system in your technology portfolio, specify its name, a descrip-

---

19 *System Actor Inventory* and *technology portfolio* are essentially the same thing, but with their equivalents for people—a "human actor inventory" is slightly less offensive than a "human portfolio." Both are equivalent to an organization chart. Also, *system actor inventory* is congruent with the Elegance methodology's references to theater and actors.

tion, and the customer experiences the system serves. Name a junior point person to contribute system details and a senior point person to contribute high-altitude perspectives on the system.

Crudely rate (high, medium, low) stakeholders' appetite to explore a change from the status quo. Rate the system's obsolescence or redundancy and the "hurdle" cost to replace it (akin to the cost of a project to replace). Name the best alternative that would reduce ongoing costs and rate the expected level of cost reductions.

A template for the System Actor Inventory is on the next page.

Skeptics are often possessive about a system because of their history and association with it. An employee with tribal knowledge sees job security in their system knowledge and subject matter expertise. They might acknowledge that a system change is valuable to customers and the company while it threatens their own comfort zone and personal job security. But as decades pass, company mergers, technology evolution, and even employee retirements force change. Skeptics must choose whether to roll with change on their own terms or be rolled by change on someone else's terms.

Certain system actors belong in your System Actor Inventory even though they reside outside your company and control and work against you. Bad actor systems are outside threats that pose financially devastating cybersecurity risks. Cybercriminals are unconventional stakeholders, but your innovation decisions and priorities should take them into account. In your asset portfolio, the System Actor Inventory is the most downstream asset where they should first appear. Some information in this template might not apply or be available, but a disciplined and empathetic methodology proactively manages for bad actors. Adhere to DevSecOps guidelines and include bad system actors in your innovation life cycle.[20]

Assign a junior employee (separate from the junior point persons cited in the Current State Inventory) to *draft* the first version of the System Actor Inventory. They will need significant help to know whom to consult as they

---

20 DevSecOps is short for development, security, and operations. It is a methodology that integrates security at every phase of innovation.

Name	Description	Customer Experiences	Junior Point Person	Senior Point Person	Appetite for Alternative (HML)	Obsolescence or Redundancy (HML)	Estimated Hurdle to Replace (HML)	Best Cost Alternative	Ongoing Cost Savings (HML)

populate the asset. That help takes the form of senior point people's informal networks within the organization. Two helpful employees turn into ten, and ten turn into twenty—all adding their perspective to the asset.

The junior and senior point people in the inventory *review* and *revise* it. A senior technology employee *approves* it. The junior employee *distributes* it to the relevant contributing and noncontributing stakeholders. Independent of projects, the junior employee should execute Five Verbs for the asset quarterly. Every project should also execute the Five Verbs as part of the Closure Report. Although maintenance of the System Actor Inventory is an excellent opportunity for a junior employee to meet dozens of coworkers, this benefit plateaus after about one year. Rotate the maintenance responsibility among junior employees.

The first version of the System Actor Inventory doesn't have an upstream asset because it doesn't wait on any other asset. The Closure Report is its ongoing parent asset because it instructs every project team to merge its future state documentation into CSD. The System Actor Inventory's downstream assets are Current State Scripts.

A System Actor Inventory shapes your culture's discipline by giving *transparency* to potentially *wasteful* technology and *costs*. The asset shapes your culture's empathy when a system accepts *rejection* by the company and *rolls out* of the technology portfolio.

Current State Scripts

Text-only lists of the sequence of actions and actors of current operations.

Process flows are standard documentation for process information. But they can be difficult to build, even for current state operations. Current State Scripts (CSS) are interim assets to ease building Current State Process Flows. CSS emulate a theater script with two columns—actor and action.

A typical business does not build CSS. CSS seem so mundane that employees might feel insulted contributing to them. The asset's high accountability and low ambiguity are so unusual, the asset can make employees nervous.

As authors Chris Zook and James Allen note in their book *Repeatability*, "Complexity is a silent killer." A script exposes the complexity of a mangled process so brazenly that it practically screams to an innovation team for help.

Scripts are conducive to exceptional detail. The left column specifies the actor. If the actor is a human, the task is manual and "hands-on." If the actor is a system, the task is automated and "hands-off." The right column specifies the action. The script is so plain that the only way to show substance, credibility, and complexity of the process is by including excellent detail.

Building a detailed script can be painstaking work. Mixing verb tenses (past, present, future) and writing in a passive voice ("the race distance is chosen") make the job harder. Instead, keep the language simple and write in the present tense. The script prevents passive voice slipping in by requiring the writer to specify the actor, thus forcing active voice.[21] Emphasis on one noun (actor) and one verb (action) makes a script look and feel robotic—a good quality because robotic actions can be automated.

Although the script format doesn't capture process decisions and branches as elegantly as process flows do, you can capture them by labeling each decision, branch, and merge in process branches. Use "if statements" generously to ease transferring decision logic to process flows (the downstream asset).

Below is a blank script template; the first column aids in numbering script steps.

Step #	Actor	Action
1		
2		
3		

---

21 Use of the active voice is an advantage scripts have over conventional process formats. Unless tightly governed and edited, business-as-usual (BAU) documents and standard operating procedures (SOPs) often contain high ambiguity and a lot of passive voice.

Skeptics don't appreciate how difficult it can be to draw process flows from scratch, that is, from meetings and emails. Doing so requires tackling content and cosmetics simultaneously. A script encourages contributors to focus on process content (in the script format) before focusing on documentation cosmetics (in a process flow format). Addressing these aspects of the content one at a time makes the work more manageable while improving the quality of both assets.

Some innovation professionals see the assignment to draft documentation as menial work. The assignment requires humility, especially for such a plain asset as CSS. But what accompanies the assignment is the opportunity to facilitate, crowdsource, synthesize, and reconcile many moving parts. CSS distill complexity into a ruthlessly simple format the team can use. The plainness of the asset makes it an excellent opportunity for new and junior employees to contribute.

The employee assigned to draft CSS will often interview subject matter experts (SMEs). If the drafter can type fast enough to capture the SME's full explanation, the conversation is easy to document. But some processes and conversations are difficult to keep up with on the keyboard. To preserve a high quality of information and keep the effort manageable, the employee can record the conversation. The script drafter can replay the recording later and transcribe at a relaxed pace. Scripting is tedious, and a replay minimizes ambiguity, exasperation, and rework.

The CSS don't purposely capture the problems of current state processes, but if contributors mention them naturally, don't ignore them and lose that valuable information. Mark the script steps related to the problem and listen for whether the biggest problems influence the Use Case Assessment or the Change Log.

Assign a junior employee to *draft* CSS and assign SMEs to *review* them. The same junior employee *revises* the scripts, and the SMEs *approve* them. The junior employee *distributes* the asset to interested noncontributing parties. When distant stakeholders know that more visually appealing

Current State Process Flows will immediately follow, they often express more interest in those than in the plain CSS.

Building CSS as a catch-up task is a form of paying back documentation debt. Build the assets as quickly as your circumstances allow, just as you would pay off financial debt. Prioritize CSS according to what projects will use them sooner rather than later. Once the assigned employees complete Five Verbs, there is no maintenance routine such as a monthly update. Updates originate within projects that impact processes. Updates to CSS reside in Future State Scripts (FSS). When a project completes, FSS are the new CSS.

The upstream assets for CSS are the Current State Inventory and the System Actor Inventory. Their downstream assets are Current State Process Flows.

CSS shape your culture's discipline. They are *easy* and improve the *speed* and *quality* of the next asset (Current State Process Flows). Their *economics* are attractive because you can assign much of the work to junior employees. CSS shape your culture's empathy, too, in that junior employees become *self-sufficient* in *listening* to the experiences and *stories* of customers and employees. For junior employees, low ambiguity—clarity—*is* empathy.

Current State Process Flows

> A visual form of current operational detail, business logic, and process branches.

To benefit from seeing your current state processes in a visual format, convert your Current State Scripts into Current State Process Flows (CSPF). Also called swimlane diagrams, CSPF show the distribution and handoffs of work, decision branches, parallel activity, and iterations. Scripts are instrumental for exhausting detail, but the process flow format is more conducive to spotting and isolating process problems. Compared to dry text, process flows are visually appealing and invite audiences to learn and engage.

A typical company has some process flows but rarely exploits their value. Don't casually build CSPF as a mere formality or afterthought. Instead, exhaust detail on paper, reconcile diverging realities, expose inefficiencies, and uncover slow and laborious work. Make it easy for other employees to learn about processes, isolate their biggest problems, and propose work to automate. CSPF help teams fix big problems by partitioning them among modestly sized projects.

Beware of employees who generously use the word *process* in conversations and emails but who are unable or unwilling to share CSPF. This hints that the process, business logic, and inefficiency are too complicated, complex, messy, embarrassing, or unethical to put on paper as a legitimate process flow. This is common, but if business logic is locked inside an employee's head, it's difficult for other employees to design, code, simplify, improve, or automate the process. Process flows provide a simple way to describe and envision automation, that is, moving tasks, logic, and decisions from a human actor's swimlane to a technology actor's swimlane.

Process flows compensate for a limitation of a standard business tool, the job description (JD). A JD's strength is its employee centricity. Its typical weakness is that it lacks employee context and is rarely *derived from* it. JDs are vulnerable to their authors (managers) writing them as a stream of consciousness. Suppose a manager is unfamiliar with the process detail of a certain operational job. In that case, the manager might draft the JD thinking about their own workload, what they want to delegate, and the work that falls through the cracks any given week. Job descriptions (centered on a single employee) are better built with context, that is, derived from CSPF (which center on the customer and team).

Process flow diagrams use a limited number of shapes. The swimlanes are the space for the activities of a single human or system actor. A box represents one activity of the actor, that is, it's a verb often followed by an object or piece of data. A diamond represents business logic for a decision (often phrased as a question) and the start of some process branching. Arrows show the sequence of work. Arrows with text show the answer to the question posed in the decision diamond.

Recall that CSS contain columns for actor and action. In drafting the CSPF, create as many swimlanes as the CSS have actors. As a default, translate every step of the script into a single box placed in the appropriate swimlane. In reviews and revisions of the CSPF, boxes might be split or merged. The translation should feel simple enough that a junior employee can draft the CSPF autonomously, armed only with skills for using the flowchart tool. Figure 8 is an example of a process flow.

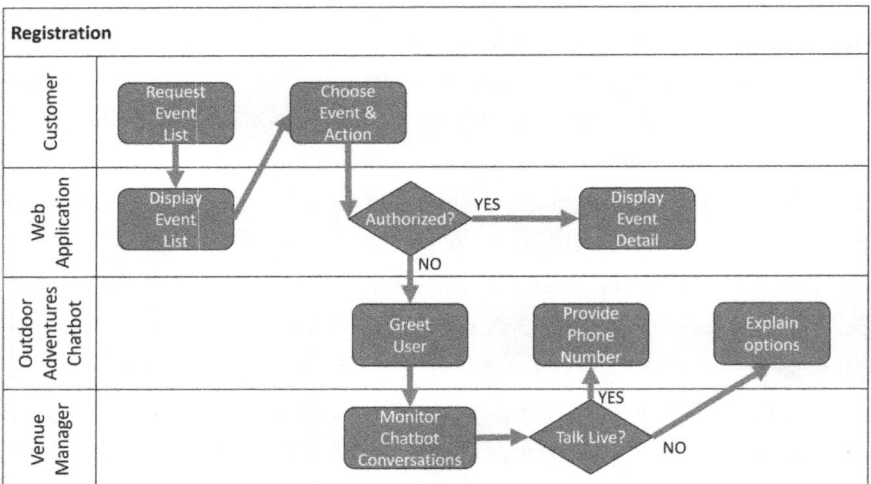

Figure 8. Current State Process Flows.

Skeptics of CSPF dislike exposing how much variability creeps into a process over months and years. Understandable, but showing the variability lowers the cost of distinguishing one-size-fits-all from customized processes. CSPF distinguish primary, secondary, and error process paths. They distinguish parallel and simultaneous processes from "either-or" process branches. They expose repetitive work worthy of automation.

Some skeptics emphasize data governance over process governance. This is equivalent to emphasizing content over context, that is, diminishing *customer* context and the *customer* experience. Context is shaped by human logic, and if a project team can't collaborate on the desired logic, the project can't automate customer context. Emphasizing data at the expense of process hinders automation and keeps operational costs high. It makes the customer experience vulnerable to whatever logic sits in employees' minds that day.

Make no mistake—a handful of data-oriented assets are vital for healthy projects, automation, and profitable innovation. Later sections describe these half-dozen assets among approximately twenty technology-oriented assets. Because of decades of software-centric methodologies and the unforgiving nature of database tools, data-oriented assets are next to impossible to neglect (unlike process flows).

A final group of skeptics might see some processes as too small and simple to justify documenting. Remind them that thirty seconds of written work saves every meeting attendee thirty minutes of talking, listening, and getting educated on a simple process, which is then still vulnerable to diverging recollections.

Assign a junior employee to *draft* CSPF, translating them from CSS. Assign subject matter experts (SMEs) to *review* them. The same junior employee *revises* them, and the SMEs *approve* them. The junior employee *distributes* CSPF to interested noncontributing parties.

Like CSS, building CSPF as a catch-up task is a form of paying back documentation debt. Build the assets as quickly as your circumstances allow, just as you would pay off a financial debt. Once the assigned employees complete Five Verbs, there is no maintenance routine such as a monthly update. Updates, which originate within projects that impact processes, reside in Future State Process Flows (FSPF). When a project completes, FSPF are the new CSPF.

The upstream assets for CSPF are CSS. The downstream asset is the Customer Experience Hierarchy. CSPF shape your culture's discipline when process flows expose the *balance* and *variability* of the operational workload.

They also isolate opportunities for *automation*. CSPF shape your culture's empathy when process flows reveal customer *points of pain*, *moments that matter*, and whether operations are out of *control*.

Customer Experience Hierarchy

The hierarchical representation of how your customers engage with you.

The current state assets explained so far are author-friendly and grounded in past work. They don't ask your innovation team to empathize with particular stakeholders. The upstream assets don't have the future in mind. The Customer Experience Hierarchy (CXH) inverts that logic.[22] It reorganizes current state operations into a customer-centric structure so future innovation has a customer-centric starting point. Customers ultimately see webpages, showrooms, and store shelves, but healthy customer experiences originate in the language and organization of the CXH asset.

A typical company has heard of "customer experience" and "the customer journey." It may have even documented a customer journey map, which captures valuable information about customers' experiences, such as their emotions. The documents include graphic design that typically makes them nothing short of beautiful. They are conducive to presentations and admiration. Their disadvantage for innovation is that they don't cry out for change and improvement. The CXH does not emphasize graphic design, so it invites changes and additions, that is, innovation. The CXH resembles a multilayered process inventory.

Each item (process) in the CXH has a title. Concise titles are best—even as short as two words, such as one verb and one object: *plan trip*, *buy*

---

22 The broad interpretation here of "customer" is an outside organization that affects your organization's finances, operations, and innovation. Conventional customers are revenue-producing. A company's regulatory, compliance, and audit customers are cost-avoiding. A company's internal customers are its employees.

food, prepare vehicle. Like any hierarchy, your CXH has tiers. You might call high-level tiers "parents" and detailed tiers "children." It's easier to use customer-centric language in the parent processes, but because child processes are behind the scenes, inevitably they are written with less customer-centric language.

The upstream, previous current state assets may or may not emphasize customers. Building an *author-centric* process inventory, an *employer-centric* solution inventory, or a product catalog is rational.[23] These assets *meet the authors and contributors where they are*. To *meet customers where they are* and optimize their experience, phrase and organize what you offer them in a customer-centric way as early as possible—in the CXH. Thoughtfully partition (modularize) processes in the hierarchy in a way that helps the customer decide what fraction of your capabilities they want to purchase. For example, the hierarchy below helps a customer eat while skating. Customers might choose only the three "planning" experiences or only the "join" experiences.

The first time you build the CXH, start from scratch. Once you've exhausted top-of-mind ideas, a cursory look at the Current State Inventory (CSI) might expose more content to add. However, don't lean on upstream current state assets because they contain significant organizational baggage you should lose. Build, arrange, and organize processes into parents and children. This fresh hierarchy, which often uncovers processes that lack Current State Scripts (CSS) and Current State Process Flows (CSPF), encourages transparency. Following is an abbreviated example of a CXH for the fictional company Outdoor Adventures:

Camp

Learn	Book, Video, In-person
Plan	Blog, Video, In-person
Join	Web
Cook	Blog, Video, Products
Sleep	Products

---

23 Solutions and products are perfect *content* to include within customer *context*. These involve and emphasize technology and data, downstream assets, and later sections of the book.

Hike

	Learn	Book, Video, In-person
	Plan	Blog, Video
	Join	Web, In-person
	Climb	Practice, Test, Products
	Walk	Products
Skate		
	Learn	Video, In-person
	Plan	Blog
	Join	Web, In-person
	Eat	Products
	Dine	Video, Partnerships, Tour
	Clean	
	Compete	

This hierarchy shows three tiers. The first tier contains three customer experiences (camp, hike, and skate). The second tier includes some customer experiences with common names (learn, plan, join), some of which are unique to the first tier (compete is unique to skate). The third tier contains a mix of communication channels (blog, video, in-person) and product references. It contains unique activities such as testing (under hike) and partnering with restaurants (under skate). Although not shown, it's easy to imagine a fourth tier.

As you build the CXH from scratch, you will exhaust top-of-mind ideas at some point. To continue adding to it, refer to the upstream CSI and CSPF and match where each document and process (respectively) fit within the CXH. You likely will find unmatched or orphaned items. For an item in your CSI or CSPF that has no place in your CXH, propose discontinuing the process (via the Change Log, Roadmap, and a project). For an item in your CXH that has no documentation, prioritize and pace building those assets among other neglected current state assets (repaying documentation debt).

Skeptics don't see the clarity of relationships and reusability that the CXH provides a business. If a business has sprawl in its documentation, it has sprawl in its operations. The CXH is a vital step (again, via the Change Log, Roadmap, and a project) in reducing operational sprawl. Even for a large business with countless moving parts, as several employees *draft, review, revise,* and *approve* the asset, the CXH shows a thoughtfully organized view of operations conducive to customer-centric innovation.

Modularity in the CXH shows where reusability is possible. In the hierarchy above, the customer experience "Join" pertains to joining a group (to camp, hike, or skate).[24] Many similarities exist in building that customer experience across the three parent customer experiences.[25]

Executing Five Verbs for the CXH typically requires several highly tenured employees with tribal knowledge. Relying on junior employees doesn't improve the speed or quality of the asset and risks "too many cooks in the kitchen." It demonstrates healthy accountability for experienced employees to shape this—and the next (last)—current state asset. Assign the tenured employee passionate about the asset to *draft* the CXH. Assign other tenured employees to *review* and *revise* the asset. Assign tenured employees with a broad view of the business to *approve* the asset. The original drafter *distributes* the asset to relevant noncontributing stakeholders.

The upstream assets for the first version of the CXH are CSPF. The ongoing parent asset is the Closure Report, which instructs the team to merge every project's future state documentation into current state documentation. The CXH's downstream asset is the Use Case Assessment.

A CXH shapes your culture's discipline by converting *sprawling,* legacy, author-centric, and employer-centric views of your business into a structured,

---

24 The hierarchy in this section shows examples of parents each with multiple children. "Join," "Learn," and "Plan" are inverse examples, that is, each child having multiple parents, a situation that is not explored here.

25 Detailed process flows comprise each customer experience in the hierarchy. High-level process flows (HLPF) bundle experiences and encompass the entire hierarchy. HLPF show whether upstream and downstream relationships are mandatory or optional. HLPF resemble the theater reference "Run of Show" and encompass the entire audience/customer experience.

svelte, *customer-centric* view poised for innovation. The CXH reorganizes *variability* and exposes *waste*. It shapes your culture's empathy as an early step toward reducing *messiness* and *integrating* many moving parts of your business. Customer empathy is the seed of innovation, and the CXH raises the tangibility and integrity of that empathy.

Use Case Assessment

> The detail and judgment that suggests which customer or employee goal is best to innovate.

The final current state asset is a self-critique—an assessment—that exposes where innovation is likely to be the most impactful. The asset is customer-centric and uses the Customer Experience Hierarchy (CXH) as a starting point. Because the assessment includes operations that take place away from the customer, the asset title uses the generic term *use case* to describe the variety of business activity within the asset. In asset form, this critique involving customer and noncustomer business activity is the Use Case Assessment (UCA). The UCA aims to identify the most compelling candidates to innovate next. It represents your final payment of documentation debt.

A typical company performs audits and critiques its operations, but an intensely transparent and methodical critique that uncovers vulnerability is more ambitious and less common.[26] It requires involving the stakeholders who are most likely to criticize. The most negative feedback is difficult to hear and distracts from responding, improving, and managing expectations and perceptions. Maximizing the usefulness of a critique and minimizing unhelpful criticism and distractions are valuable practices, and skills and tools for such capabilities don't aim for objectivity. They welcome and *need*

---

26 And an earlier asset, the Voice of the Customer and Seller, represents a less prescriptive and less detailed critique.

subjectivity and judgment. For the best critique, your team must be humble, vulnerable, and methodical. The UCA is the asset that scrutinizes culture traits for discipline and empathy.

The UCA asks stakeholders to judge a half-dozen aspects related to discipline. One aspect is the number and nature of *actors* executing the use case. This probes whether the number might be too high or too low and whether the use case is overly manual or automated. Consider the *steps* to complete the use case. A precise number isn't necessary—the goal is to probe whether the use case is overly complicated or complex. Judge the number of *handoffs* among actors because these require synchronization and increase risk. Assess the *duration* because a long duration often signals a bottleneck. Assess the *effort* because laboriousness is expensive. And, finally, judge *fragility* because any fragile process is prone to break down, halting the entire process. Innovating fragile processes into robust processes reduces this risk.

Judging these aspects of process discipline does not have to be scientific or precise. A Stoplight Report (Red, Yellow, Green) is suitable to differentiate candidates for innovation.

In contrast to the pragmatism of process discipline, the UCA also captures stakeholder sentiment—empathy—for change. Stakeholders judge the desire to change, the perceived or expected effort to change, and the expected impact of the change. Instead of stoplight colors, rate these as high, medium, or low to differentiate candidates for innovation.

One might expect the three stakeholder empathy ratings to be correlated; for example, high impact accompanies strong desire, or high effort accompanies low desire. But correlation isn't certain. The assessment captures all three ratings because any of them can make a use case a better or worse candidate to innovate next.

Although not reflected in the template below, a more detailed, empathetic approach to the UCA uses the concept of the *Eight Ds* (dull, dirty, dangerous, difficult, demanding, demeaning, delicate, and dear). Having stakeholders issue a Stoplight Report on these traits is a deeply empathetic approach to uncovering innovation candidates.

Altogether, the UCA contains six judgments about process discipline and three judgments related to stakeholder empathy. Again, one might expect ratings between the two categories to be correlated. For example, low desire and low impact accompany Green ratings, or strong desire and high impact accompany Red ratings. But, again, correlation isn't certain. Contributors and stakeholders of the UCA should understand consistencies and inconsistencies because both can make a use case a better or worse candidate to innovate next.

The first step to build—*draft*—the UCA is for a junior employee to translate every child in the Customer Experience Hierarchy to the UCA as individual use cases. To establish accountability for each use case, identify two points of contact—a junior point person (JPP) for detailed, local matters and a senior point person (SPP) for high-level, global matters. Describe every use case's frequency (hourly, daily, weekly, monthly) because high-frequency use cases have a magnified impact compared to low-frequency use cases. Prioritize high-frequency use cases in populating the rest of the UCA and judging candidates for innovation.

To populate the remaining nine pieces of information for each use case, the drafter leans heavily on each SPP. Typically, the SPP seeks opinions and ratings from many relevant stakeholders. The JPP can handle scheduling and logistics for this *review* and *revise* of the asset for their fraction of use cases.

Every SPP contributes their respective information to the junior employee who reviews and revises *the entire asset*. A senior employee *approves* the asset as a whole and *distributes* it to stakeholders who need to understand the judgments that likely shape innovation priorities.

In building the asset, the altitude of a use case can feel too high or too low,[27] or a use case's optimal granularity may not be apparent. The optimal altitude for a use case is low enough to clarify ownership (of the use case and the row on the asset) and reconcile disagreements that are GETMO (good enough to move on) within an assessment. The UCA doesn't dictate what to innovate next, but it exposes the best candidates.

---

27 Splitting and merging use cases in the UCA justifies revising the upstream asset, the CXH.

Below is an example of a UCA for Outdoor Adventures.

Use Case Title	Junior Point Person	Senior Point Person	Frequency	Actor List	Steps	Handoffs	Duration	Effort	Fragility	Desire to Change	Effort to Change	Impact of Change
Learn			Monthly	Green	Green	Green	Green	Green	Green	L	M	H
Plan			Quarterly	Green	Green	Green	Green	Green	Red	M	M	M
Join			Annually	Green	Green	Green	Green	Green	Green	L	L	L
Climb			Monthly	Green	Green	Green	Green	Green	Green	0	0	0
Walk			Weekly	Green	Green	Green	Green	Green	Green	0	0	0
Dine			Monthly	Red	Green	Green	Yellow	Green	Green	L	L	L
Clean			Annually	Green	Green	Green	Green	Red	Green	H	H	L

This example shows four candidates to innovate next: the "Dine" use case, because of the number and nature of actors; the "Clean" use case, because of the level of effort to clean skates; the "Plan" use case, because planning feels like a fragile process; and, although no one is clamoring to scrutinize the "Learn" use case, the team feels innovating it would have high impact.

A skeptic of the UCA (predictably an SPP) is an innovation "contrarian" who lacks the humility, vulnerability, and incentives for participating in a transparent, methodical critique of the business's current state. One problem for skeptics is that their disinterest ensures they are among the least-informed employees about existing problems, whereas their colleagues are aware, informed, and ready to address known problems. Skeptics' other problem is their propensity to rate operational health as overly rosy. This bias toward rosy health ratings guarantees innovation neglect.

Disagreement about decisions *within* the UCA is normal and healthy. Disagreement about *whether* to document a UCA is unhealthy. The lack of a UCA places contrarians and their teams at a competitive and collaborative disadvantage.

Another way to use the UCA is to consider use cases your company does not yet offer. Employees might be aware, informed, and ready to address known problems outside your company (these appear in your Market Forces Matrix). If your company can create (i.e., innovate) and manage a customer experience better than other companies, you stand to increase your scope, impact, and revenue.

Because the UCA depends on the CXH (an upstream asset), execute Five Verbs for a first build of the UCA in the shadow of the CXH. However, information in the UCA resembles and is a bridge to the information in the Change Log (a downstream asset). Diligent maintenance of the UCA emulates and foreshadows maintenance of the Change Log, that is, it's done monthly (preceding reviewing and revising the Change Log).

The UCA is among the most threatening assets to contrarians who are sensitive to critique and scrutiny. Disagreements are understandable and common. As always, the employee assigned to *approve* the UCA serves as a tiebreaker for disagreements, and they also determine the transparency of the asset. If you experience an impasse with a contrarian within the UCA, elaborate on the innovation candidate in the Change Log and break deadlocks in that asset.

Upstream assets of the UCA include the Scorecard, Current State Process Flows, CXH, and the Market Forces Matrix. The downstream assets include the Change Log, Customer Lifecycle Model, and Use Case Definitions.

The UCA shapes your culture's discipline by routinely judging the *quality* of every use case and exposing high-*quality* candidates for innovation. The UCA shapes your culture's empathy by *vulnerably* rating stakeholder *sentiment* for change, separately from process discipline. The UCA builds and maintains institutional skills for giving and receiving *rejection*.

A team that completes all current state assets and maintains a UCA has no *documentation debt*. A team that habitually maintains all project-independent assets is on an elegant path to having no *Methodology Debt*.

PROJECT-SPECIFIC ASSETS

Of all the things I've done, the most vital is coordinating the talents of those who work for us and pointing them toward a certain goal.

~ Walt Disney (1901–1966), American animator, film producer, and entrepreneur

An innovation team with a complete project-independent asset portfolio is set up to succeed with any project it pursues. Within every project, healthy innovation teams maintain discipline and empathy in building the right process, people, and technology assets.

A typical project team is familiar with these assets but treats some casually. A typical project team is also familiar with the counterproductive documents explained in Appendix B. Although a casual approach and counterproductive documents are better than no documentation, they distract from building the right assets and dilute teamwork, undermining a culture of discipline and empathy. Committing these assets to muscle memory maximizes the value of your teamwork.

Some skeptics want to cut corners or race ahead to the work they are good at, such as design or code work. This is understandable, but it causes ambiguity, errors, and rework. For project-independent assets, rhythm and frequency matter more than sequence. For project-specific assets, sequence matters, and the ideal frequency to complete them is once—getting them right the first time. Whereas the Agile methodology embraces being iterative, the Elegance methodology emphasizes small projects, completed the first time to the level of GETMO (not perfection) with attention to detail, with additional distinct projects in the queue.

As use of these assets becomes routine, junior employees can do more of this heads-down work, freeing senior employees to do more heads-up work. Small and simple projects still complete these assets but with much less effort, duration, and cost. Junior employees can handle the mechanics of project-specific assets while senior employees explore additional innovation opportunities according to their expertise and style.

These project-specific assets are not technology-oriented but can be categorized as project management, process, and people assets. Completing these assets before touching technology assets is another way that a people-centric methodology distinguishes itself from software-centric methodologies. The Elegance methodology keeps technology, data, and content in service of people, processes, and context.

Assets in the three categories of project management, process, and people have a large "blast radius," that is, if an error originating in these assets survives a long time, the error can have an outsized negative impact on rework, schedule delays, and the project's value proposition. In contrast, errors originating in a technology asset have a relatively small impact on rework, delays, and the project's value proposition. A typical omission in technology assets is not crippling in terms of duration, effort, or cost, but a typical omission in these three categories of project-specific assets can be crippling. For decades, organizations have been conditioned to pay close attention to detail in technology assets, but not in project management, process, and people assets.

Project-specific assets shape your culture's *discipline* by keeping the team's *quality* high and *waste* low. The stable asset portfolio makes *vigilance* easy. Project-specific assets shape your culture's *empathy*. Whereas project-independent assets emphasize frequency and rhythm, project-specific assets demand heightened alertness for durations, deadlines, and schedule *visibility*. These assets unambiguously assign work during the project and *set expectations* for how stakeholders' experiences will be different after the project. Every project disrupts the status quo, and these assets oversee disrupted accountability with sensitivity to schedule in the spirit of *bang-the-table* leadership. Their value—and the damage in their absence—applies to every project so that innovation teams can avoid *reinventing the wheel*. Project-specific assets "name names" and shape *simple* views of the past, present, and future. The asset group is vital for *minimizing surprises* and building *trust* for employees, customers, and stakeholders.

Project Management

A project is forgiving at the beginning or the end—never both.

~ Unattributed

The first questions to answer about every project are captured in a handful of assets categorized as project management (PM). The PM assets, including Project Charter and Business Case, Project Plan, Workstream Status Report, and Traceability Matrices, govern work on all other project-specific assets—process, people, and technology.

A typical project team builds some combination of PM assets, but the rigor of the tools, content, and administration varies widely. Variability and unpredictability of PM assets are forms of "reinventing the wheel" and a major source of poor *discipline* and *empathy*. A company reduces its PM costs by standardizing content and administration rigor and by striving for predictability and minimal surprises.

PM assets shape your culture's discipline because repeated, rhythmic stakeholder alignment on them is a sign of high *quality* and respectable *speed*. Standardizing PM assets minimizes *waste*, discourages *variability*, and makes PM *straightforward*.

PM assets shape your culture's empathy. Their *transparency* (to stakeholders) and *visibility* (weeks into the future) *set the table* for employees, customers, and outside stakeholders to be *confident* in their collective success.

Project Charter and Business Case

> A gathering of vital information
> about a project at its inception.

The first step for any single project is to build a Project Charter. The charter's goal is to align all early expectations of a project. A similar asset—a Business Case—can contain similar information. Although an innovation team might partition early project-specific information, this asset presents the Project Charter and Business Case together.

The typical project team documents some form of a Project Charter and Business Case, but the emphasis within these assets varies from company to company, team to team, and project to project. For example, a project team could include financial projections or organizational changes. The Elegance methodology's Project Charter and Business Case template excludes these topics because they are difficult to standardize. Rarely are they a root cause of project failure; guidance on such topics is widely available elsewhere. The Project Charter and Business Case template below includes the items teams must agree on first to minimize negative surprises and project failure later.

Following a rigorous template is vital because the most common project trouble originates in neglect of the Project Charter. But note that rigor is different from large project scope. To keep risk low, keep the scope of every project modest. The most significant indicators of project scope and size are a few quantities in the Project Charter: the number of use cases, impacted underlying actors, and process steps.

Rigor is also different from building one comprehensive project document. Don't treat the Project Charter and Business Case as a dumping ground for expectations more appropriately placed in process, people, or technology assets. If the following template contains information that doesn't resonate with or apply to your project, state that in the charter. Collaborate on populating the information that does resonate with your team and stakeholders.

Section	Description
Title	*Project Name:* _____
Executive Summary	Consider just three statements that imitate the ORS Report asset. State the relevant *observation*, the *reaction* from the key stakeholder who cares, and the *suggestion* to improve the stakeholder experience.
Problem Statement(s) or Points of Pain	Describe and explain the negative experience.
Aspirations	Insert the primary stakeholder's instinctive answer for what they hope for. Finish the sentence, "It would be great if _____."
History /Project Context	Share any past events the project team should know about as the project begins.
Origin or Root Cause(s)	State this to avoid misperceptions or mislabeling "symptoms" of the unacceptable status quo.
Scope Description or Project Objective	Elaborate on the aspiration (stated above). List the intended customer experiences and stakeholders within the boundaries of scope.
Success Factors Lessons Learned Premortem	List relevant project qualities learned from previous projects. Don't re-create an entire Lessons Learned exercise but include three to five points uniquely relevant to this project.
Impacted Use Cases	List use cases or a hierarchy of what the team expects the project to impact (new, revised, discontinued).
Out-of-Scope Specs	List customer experiences, use cases, and stakeholders just outside the project boundaries.
Sizing Defense	Explain why a larger or smaller project scope is less ideal. List use cases that just "made the cut" or just "missed the cut."
Relevant Metrics	List leading metrics—recent historical values and target values. List lagging metrics—recent historical values and target values.

Description of Financial Impact	Describe the expectation of revenue rise or cost reduction.
Cost of Delay	List relevant revenues and costs of doing nothing. List relevant revenues and costs of investing (expected). Calculate the difference between these two amounts.
Dependencies on Other Projects or Competition of Investments	List concurrent projects with higher priority. List concurrent projects with lower priority. List interdependent projects.
Sponsor(s)	Name the most senior person with visibility to the project who wants the project to succeed. They are accountable for ensuring the investment adheres to the Value Proposition and Benefits Realization. They are the final point of escalation and the tiebreaker for team disputes. They are empowered to suspend or cancel the project.
Core Team	Name employees who spend five or more hours per week on the project, appear a lot in the Project Plan, and participate in weekly status meetings.
Extended Team	Name employees who spend five or fewer hours per week on the project and appear sparingly in the Project Plan.
Extended Stakeholders	Name employees, customers, or other outsiders who solely monitor progress.
Staffing Needs	List special, unusual, or surprise personnel needs.
Impacted Systems Actors	Name the systems that change their role as part of the project, i.e., new, modified, or expired.
Assumptions	List statements that, if true, allow certain work to be unnecessary; and if determined false in the future, signal that the team has new work to do, stakeholders could change, and the schedule will undoubtedly change or be delayed.
Communication / Change Management Approach	Name the approach or the frequency of various meetings. Name the approach or frequency of formal project updates or checkpoints.

It's common for other information to arise at the start of a project, such as detailed functional requirements, technical requirements, training expectations, or a go-live sequence. Recording this information in the Project Charter is premature, but if it helps the team move forward, capture and label it as preliminary information and have the appropriate asset host the information later.

Some skeptics of the Project Charter and Business Case are impatient with this information gathering. They want to move on to other project decisions and other assets. But thoughtful work early in a project reduces the original stakeholders' confusion later and decreases onboarding work with stakeholders who join the project after the team approves the charter.

Other skeptics feel it's rigid to set all these expectations. *Rigor* doesn't mean *rigid*. Rigor is low ambiguity and high alignment. For the Project Charter and for every asset—if the team learns something that justifies revising an asset, a safe, healthy team revises the asset and propagates that change through every impacted asset. Transparency of critical project information increases the safety of changes, *reducing* rigidity.

Projects are forgiving at the beginning or end, but never both. As a project progresses, *code* is unforgiving. When a project is finished, *customers* can be unforgiving. This is painful for a project team that plays things loose. A healthy project team is rigorous and unforgiving with its work *early* in the project. It anticipates where stakeholders might have diverging expectations and pursues alignment to minimize surprises and negative stakeholder reactions later. When a team systematically pursues transparency and converging expectations, the late phases of a project are only vulnerable to small mistakes that resemble sampling errors. Such rigorous projects can fix and forgive work where rework or delays are minor and contained. But a *forgiving* conclusion to a project requires an *unforgiving* Project Charter.

Upstream assets include the Roadmap, Change Log, Use Case Assessment, and Scorecard. Downstream assets include the Customer Lifecycle Model and Awareness Blast.

Because the Project Charter and Business Case is such a consequential asset, executing Five Verbs deserves some special vigilance. A junior employee can *draft* the charter but must gather information from experienced employees and stakeholders with unusually high inclusion and vulnerability. The number of stakeholders who *review* and *revise* the asset and the diversity of their expectations can be high. Reviewers must include the senior point person for every affected use case. The project sponsor and other relevant senior employees who *approve* the charter might have a difficult job reconciling, aligning, and converging on the charter's content. The junior employee *distributes* the charter to relevant noncontributing stakeholders, although a mature team is cautious about its audience when the charter contains sensitive information.

The Project Charter shapes your culture's discipline. A modest scope avoids jeopardizing the *economic* value proposition of the project. The asset minimizes the *variability* of previously diverging expectations and keeps *speed* high when additional stakeholders need to learn about the project. The charter shapes your culture's empathy by minimizing the *messiness* of diverging expectations and unspoken assumptions. A rigorous charter requires *inclusivity* and *vulnerability* to populate. It *blends* diverse stakeholders' goals into a singular vision for a single project.

Project Plan

A detailed representation of a project's phases, assets, activities, dependencies, assignments, schedule, and progress.

The Project Plan assigns, synchronizes, paces, and tracks the completion of innovation work. The Project Plan is instrumental in shaping a healthy employee experience and setting expectations for a realistic completion date all while minimizing negative surprises related to assignments and schedule.

A typical project team builds Project Plans; however, the tools, content, and attention to detail vary widely. Many project teams use a spreadsheet tool meant for sophisticated math. Several tools exist whose sole purpose is planning, synchronizing, and tracking the completion of teamwork. These tools reduce the upfront and ongoing labor of building and maintaining a high-quality project schedule.

Regardless of tool choice, a poor Project Plan is a common weakness that puts teams at a disadvantage. A poor Project Plan contains sprawling, ambiguous language that attempts to manage disposable, noncollaborative work. A good Project Plan, in contrast, contains unambiguous language that governs collaboration on durable work. Whereas a poor Project Plan is a simplistic, casually formed, reinventing-the-wheel to-do list that hides sequence, dependencies, assignments, and duration of work, a good Project Plan regulates sequence, dependencies, assignments, duration, and percent complete and organizes these in a way that applies to every project.

What qualifies as durable work is documentation that withstands the rigor of Five Verbs. Without a draft, work isn't written down, and employees remember decisions differently, if at all. Without reviews and revisions, work risks silos and the exclusion of important contributors. Without approval, work lacks oversight and accountability. Without distribution, work (figuratively or literally) sits on a shelf, unused by project colleagues, and proving of no valuable to other stakeholders.

Work not governed by Five Verbs lacks collaboration and integrity. Trying to use that work again by, for example, retrieving emails and remembering meeting highlights, is expensive and laborious, rendering it forgettable and disposable. A Project Plan doesn't need to *forbid* other work, which would be laborious and petty. Nor does a Project Plan need to exhaust every motion of an employee since doing that has no long-term value and is impossible to standardize.

Not only is durable work itself easy to use again, but also the simplicity of the Project Plan helps recall who did the work and when. A junior employee can inspect a Project Plan to see asset names, assignments, durations, dates, and percent complete. Other language in a Project Plan increases ambiguity and laboriousness in determining who did the work and when.

Another benefit of a Project Plan that leverages Five Verbs is that simple dependencies are visualized and can be used to control the expected duration of the entire project, that is, a Project Plan delineates the project's "critical path." The formal definition of *critical path* is "the sequence of activities that represents the longest path through a project, which determines the short-est possible duration." Changing the duration of work on the critical path affects the entire project's duration. Changing the duration of work off the critical path does not impact the entire project's duration. A Project Plan with complicated dependencies complicates managing a project's critical path, but a Project Plan with simple dependencies helps the team see where project delays might arise.

Organizing work that applies to every project requires that the Project Plan serve the three categories of process, people, and technology. Within those categories, specific assets record the answers to questions relevant to every innovation team.

Figure 9 shows a slice of a Project Plan (showing three of thirty project-specific assets) displaying all the ingredients: assets, five verbs, dependencies, assignments, duration estimates, and percent complete. The simplicity of the plan's building blocks creates an elegant Gantt chart (the bars on the right).

Because the Project Plan is so standard and straightforward, a junior employee can maintain it. They receive new information through the formal channels of status meetings and Individual Status Reports and the informal channels of conversations and emails.

Some information in the Project Plan rarely changes during a project, and some changes weekly. Assets and tasks (verbs) rarely change. Dependencies are subjective—changes are rare and minor. Assignments and duration estimates *can* change often. Percent complete values *should* change (rise) every week.

A Project Plan based strictly on assets and Five Verbs has a few different skeptics. Some feel a plan that contains thirty assets and five verbs is overkill. They prefer to manage fewer assets and organize the work less formally. But eliminating assets and verbs hurts transparency (of information), visibility (of schedule and critical path), and accountability (of assignments) and increases the cost of information sharing.

Other skeptics feel that Five Verbs is limiting. They want to plan according to their comfort zones of familiar decisions and typical meetings.

	Task	Predecessors	Assigned To	Duration	Start Date	End Date	Status
1	Project Alpha			27d	01/01/30	02/06/30	
2	Project Charter			10d	01/01/30	01/14/30	
3	Draft		Avery	4d	01/01/30	01/04/30	Complete
4	Review & Revise	3	Bailey,Casey	3d	01/07/30	01/09/30	Complete
5	Approve	4	Dakota	2d	01/10/30	01/11/30	In Progress
6	Distribute	5	Avery	1d	01/14/30	01/14/30	
7	Training Approach	2		6d	01/15/30	01/22/30	
8	Draft		Easton	1d	01/15/30	01/15/30	
9	Review & Revise	8	Finley, Gabriel	3d	01/16/30	01/18/30	
10	Approve	9	Harper	1d	01/21/30	01/21/30	
11	Distribute	10	Jordan	1d	01/22/30	01/22/30	
12	Training Materials	7		11d	01/23/30	02/06/30	
13	Draft		Kerry	3d	01/23/30	01/25/30	
14	Review & Revise	13	Lou, Maxine	2d	01/28/30	01/29/30	
15	Review & Revise	14	Noel, Oakley	3d	01/30/30	02/01/30	
16	Approve	15	Paris	2d	02/04/30	02/05/30	
17	Distribute	16	Kerry	1d	02/06/30	02/06/30	

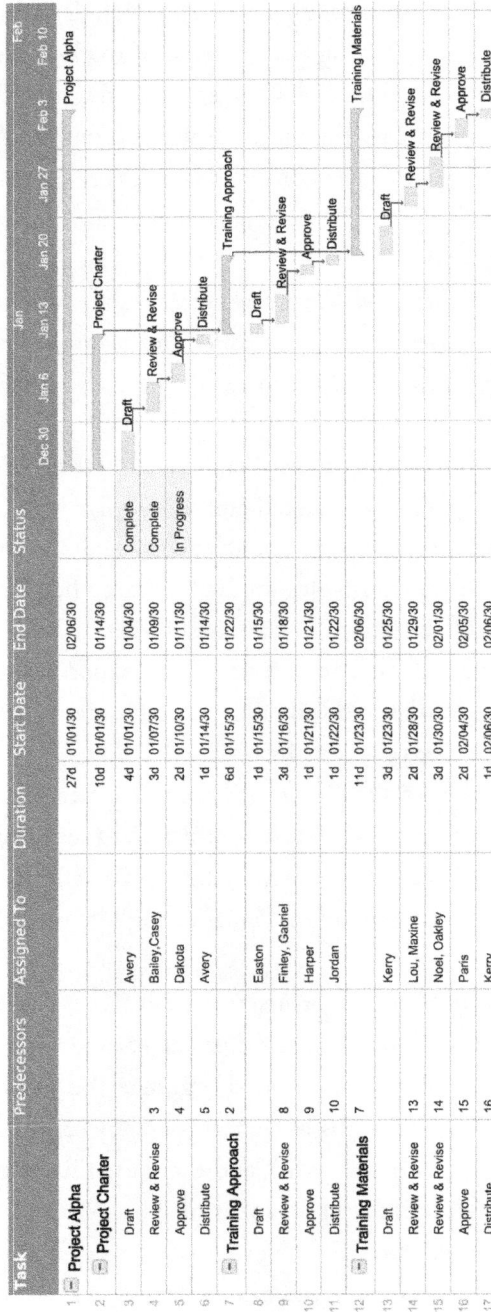

Figure 9. Slice of Project Plan.

They prefer language freedom, spontaneity, and Verb Sprawl. Adding tasks unrelated to documentation, collaboration, approval, or information sharing isn't crippling, but doing so micromanages work that is of low value, interest, and risk—that is to say, *disposable work*. When employees want their manager to know about work that doesn't benefit their colleagues, they can claim the credit on their Individual Status Report. Too much freedom and Verb Sprawl in a Project Plan causes the Project Plan to be less useful and, at its worst, unintelligible. Standardization of a Project Plan makes building, maintaining, and absorbing the plan inviting and an elegant exercise. Administering a Project Plan with clear assets, verbs, assignments, and durations is simple and inexpensive.

Another type of skeptic assumes the worst about the project schedule. They equate a documentation-heavy Project Plan with a year-long legacy Waterfall project. They ignore how current state assets help keep scope small and projects short. Appendix C shows a twelve-week Project Plan for a team to use as a starting point and extend and compress as they see fit.

A final type of skeptic worries about sandbagging, that is, making bloated estimates or estimating in bad faith. Legacy approaches to planning might harbor incentives for and tolerate inflated estimates, but this plan's simplicity and visibility weeks into the future replace those incentives with the risk of embarrassment that, weeks in advance, someone needs twice (for example) as long as others to do the same work. Once an overly cautious team member feels safe and familiar with the team's rhythm, visibility and peer pressure discourage bad-faith estimates.

An elegant Project Plan populates 120 (five verbs on four rows for thirty assets, so 4 × 30) assignments and duration estimates. Building the plan the first time is time-consuming. However, planning subsequent projects is much less time-consuming, and the "default" Project Plan improves with every project it serves. Every project can have an elegant Project Plan with low marginal cost.

Five Verbs applies to the Project Plan asset. A junior employee serving as project manager *drafts* the Project Plan by populating it with thirty assets, five verbs, and dependencies.[28] Core team members (as specified in the Project Charter) *review* and *revise* the Project Plan with assignments and duration estimates. The sponsor *approves* the Project Plan. The project manager *distributes* the Project Plan to the core team.

Maintenance of the Project Plan is a weekly routine. Core team members push status information to the project manager.[29] In a typical week, the project manager *revises* durations (when work runs ahead or behind schedule) and increases the percent complete for each task. A week with more surprises *revises* dependencies or assignments.[30] Weekly *approval* takes the form of the project manager *distributing* the Project Plan to core team members.

Any poor project health takes form in the Project Plan as a delay of an in-flight asset, that is, a duration estimate increases repeatedly. The generic remedy is to get others involved. For example, if the draft is delayed, have reviewers and revisers take over the asset. If reviewers and revisers cannot converge, have the approver take over the asset. Change assignments in the Project Plan to reflect formal changes. A few scenarios justify escalation to the project sponsor: repeated dependency changes, unfilled assignments, rework, and delays that postpone the project completion date by a week or more.

Because a Project Plan governs all thirty project-specific assets, it is effectively a parent asset to all of them. But the impact those thirty assets have on the Project Plan is limited to their percentage of completion as it rises from 0 to 100. Whether best described as upstream or child assets of the Project Plan, several (mostly project-independent) assets reliably or possibly influence the Project Plan weekly—especially assignments and duration estimates: Project Charter, Lessons Learned, Approachability Menu, Parking Lot, Crisis Resolution Checklist, Individual Status Reports, I Like I Wish I

---

28 This work alone, without assignments and duration estimates, contains the ingredients of the Project Plan that apply to every project, minimizing reinventing the wheel. The next project only needs to revise assignments and duration estimates.

29 When the project manager has to first "pull" status from every team member, this doubles the cost of weekly status reporting.

30 Revisions might justify a conventional weekly status meeting.

Hope I Wonder reports, and Workload Reports. Downstream assets include the Roadmap and Workstream Status Report.

The Project Plan shapes your culture's discipline. The strict boundaries of thirty assets and five verbs *ease* managing the project, limit *variability*, and minimize *waste*. Weekly increases in percent complete impose *vigilance*. Revising durations and assignments shows *elasticity*.

The Project Plan shapes your culture's empathy. The *stability* and *confidence* of having thirty assets and five verbs prevent *reinventing the wheel*. Dependencies *synchronize* work. Detailed duration estimates *pace* the work. Assignments show team members' names, demonstrating *inclusivity*. As the percentages of completion *rhythmically* reach 100 percent, *trust* grows.

Workstream Status Report

> A project's report on objectives, general schedule, primary accomplishments from the past week, and goals for the upcoming week.

Just as individual employees generate a weekly status report, diligent project managers generate a weekly Workstream Status Report for their projects. The report shares and reinforces the project's most time-sensitive information with the extended stakeholder population.

A typical project team generates status reports, but the reports are often casual, with low standardization and variable rigor. Some status reports are minimalist, that is, aiming to minimize what they share with their audience. Low transparency and high ambiguity commonly lead to poor alignment, poor accountability, and negative surprises. A standardized and unambiguous status report is engaging, maintaining trust and alignment. Such a status report uncovers diverging expectations and perceptions early and eases negative news. It's a form of receipt—evidence of alignment on certain project information at a specific time.

A Workstream Status Report contains the project's objective and success factors. The project manager gets this information from the Project Charter. Because some stakeholders don't regularly review the charter and don't always remember the objectives and success factors, the Workstream Status Report reminds them weekly. Usually, this information doesn't change during the project.

The next information on a Workstream Status Report resembles information found on an Individual Status Report but applied to the project. On a regular week, list the accomplishments from the past week. On a slow week, list any newsworthy information that represents the project's activities, circumstances, or barriers. Next, list goals for the next week. On a slow week, list any plans or circumstances that shape the next week's progress.

Next, the report shows a schedule, that is, a list of milestone dates. A high-level schedule lists expected start and completion dates for phases of work.[31] A detailed schedule lists these dates for every asset. A hybrid status report shows *assets* for an in-flight phase of work and shows *phase only* when no work is in flight for it. For example, early in a project, there's no work on technology assets, so Design, Build, Test, and Deploy are listed as phases and no individual assets are named. Late in a project, there's no in-flight work on process assets, so Future State Process is listed as a phase, and specific assets are not. Each row also reserves a space for any comments the project manager or the core team want to share with the audience of the status report.

The project manager gets dates from the Project Plan. The start date on the status report reflects the start date of drafting the first asset. The completion date reflects the completion date of distributing the last asset. The project manager also gets the percentage of completion from the Project Plan. Most planning tools provide a calculated value for phases and assets (when they don't, an approximation is suitable).

---

31 People phases of work include Awareness, Desire, Knowledge, Ability, and Reinforcement. The single process phase of work is Future State, and technology phases are Design, Build, Test, and Deploy.

Week to week, the project manager's primary input for the status report is the Project Plan. The project manager phrases each bullet point using three ingredients (in reverse order of how they are listed here): (1) the asset in flight, (2) the verb in flight, and (3) whether the assigned employees started, completed, or continued the verb. Examples for next week might include:

- Continue review and revise Project Charter

- Complete draft of code

- Start approving Test Scripts

When considering the status of items from last week, keep references to *start* and *continue* and drop any reference to *complete*. This example shows the three tasks above now complete.

- Reviewed and revised Project Charter

- Drafted Code

- Approved Test Scripts

There are two situations that justify simplifying the language and diverging from these rules. First, if start and complete occur within a single reporting period, dropping both might still be coherent (see training materials example below). If a team completes all Five Verbs within a single reporting period, dropping the verbs might still be coherent (see test execution example below).

In the spirit of an agreement factory building your asset portfolio, writing the bullet points should feel mechanical, robotic, and not creative. Creativity would expose Verb Sprawl and extraneous noncollaborative work.

Finally, append a Stoplight Report to convey health-at-a-glance. The following page shows an example of a Workstream Status Report for the fictional company Outdoor Adventures for June 5.

Skeptics prefer less transparency with extended stakeholders, especially if the project has problems or delays. Some skeptics are unnerved by the simplicity of the combination of the Project Plan and status reports. Yet

Project Objective

To build and deploy the capability to host and manage skating competitions

Success Factors

- Learn from competitors and previous races.
- Leverage third parties as much as possible.
- Avoid geographical sprawl.

Last Week Highlights

- Approved and distributed
- Training Approach
- Approved and migrated Code
- Approved and distributed Test Plan

Next Week Plans

- Draft, Review, Revise, Approve training materials[32]
- Start and complete Test Execution and Actual Results[33]

Phase / Asset	Commentary	Target Start Date	Target Completion Date	% Complete	R/Y/G
Charter		May 1	May 5	100%	
Future State Scripts		May 8	May 12	100%	
Future State Process Flows		May 15	May 19	100%	
Design and Test Approach		May 22	May 26	100%	
Build and Test Plan, Training Approach		May 29	June 2	100%	
Test Execution, Training Materials		June 5	June 9	5%	Yellow
Training Delivery		June 12	June 16	0%	Green
Deployment and Go-Live		June 19	June 23	0%	

32 Omitting reference to any of start, continue, and complete conveys that all of them apply within a single reporting period.

33 Omitting reference to any of the Five Verbs conveys that all of them apply within a single reporting period.

simplicity and transparency are execution's two best friends and are the goals of the project management assets.

Other skeptics might feel that only some projects impact all thirty project-specific assets. When this happens, the Project Charter should record the deleted assets in the out-of-scope section. The status report should formalize removal decisions as another form of simplicity and transparency for stakeholders.

Five Verbs applies to the Workstream Status Report. The project manager *drafts* it using the just-updated Project Plan (the sole upstream asset of the status report). If any ambiguity exists, the project manager *reviews* and *revises* the Project Plan, status report, or both with team members who can clarify the uncertainty.[34] The project manager effectively *approves* the status report by *distributing* it to the extended team or stakeholders (as defined in the Project Charter). Afterward, if a stakeholder has questions that result in revising the status report. If the change is time sensitive, the project manager should redistribute the status report before the next report.

The Workstream Status Report exposes poor project health as a delay of an in-flight asset. Examples include when "Last Week Highlights" report that a task "continued" in consecutive weeks, when percent complete values stall before reaching 100 percent, or when target completion dates are postponed. The audience of the status report is usually larger than the audience of the Project Plan. In the wake of a delay, vigilant project sponsors and asset approvers are ready to manage perceptions and expectations about progress.

The status report shapes your culture's discipline by managing and reporting on a short leash, which keeps the project's *economics* intact. Translating the Project Plan to the status report is automatic and practically *automated*. The status report shapes your culture's empathy because a junior employee *orchestrates* this simple weekly routine that resembles team *practice, rehearsal,* and *self-sufficiency.*

---

34 This situation justifies a conventional weekly status meeting.

Traceability Matrices

> The mapping of detail between
> two assets to verify two-way
> accountability of specifications.

Traceability Matrices ensure that everything a team intends to build *is built* and everything a team builds *is intended*. The matrices aim to minimize waste, omissions, and sneaky additions to innovation work.

A typical company has some orphans—it builds something out of the blue (known as "gold-plating") or intends to build something but then doesn't ("forgotten"). The three common scenarios of gold-plating include when a project starts as a surprise to some (it appears mysteriously on the Roadmap), when a database is changed (it appears mysteriously in the Data Model), or when someone informally asks for a change (that mysteriously appears in Code). The three common and opposite scenarios of forgotten intentions include Project Charter content that a team forgets about, Future State Process Flows (FSPF) content that a team forgets about, and content that should appear in a Test Script or training materials but doesn't. The following table identifies the upstream and downstream assets of each of these six useful Traceability Matrices:

Upstream Asset	Downstream Asset	Orphan Type
Change Log	Roadmap	Gold-plating
FSPF	Data Model	Gold-plating
FSPF	Code	Gold-plating
Project Charter	FSPF	Forgotten
FSPF	Test Script	Forgotten
FSPF	Training Materials	Forgotten

Here is an example of building a Traceability Matrix for innovating how customers plan a camping trip. FSPF are the upstream assets, and Test Scripts are the downstream assets. The first step is to list every specification in the *upstream* asset (FSPF) that you believe should appear in the *downstream* asset (Test Script).

1. Specify camper profile (age, gender, physical limitations).

2. Specify camper preferences (region, climate, terrain).

3. Specify emergency contact.

4. Choose a campsite.

5. Choose dates.

6. Choose transportation to/from the campsite.

7. Explain the policy for changes and cancellations.

The second step is to list every specification in the *downstream* asset (Test Script) that you believe should originate in the *upstream* asset (FSPF). Test Script

A. Organizer receives camper's profile (age, gender, physical limitations, dietary restrictions).

B. Organizer receives camper preferences (climate, terrain).

C. Organizer calls the emergency contact.

D. System presents campsite options and available dates.

E. System proposes alternate campsite based on camper age over sixty.

F. System captures camper transportation.

G. Camper accepts policy for changes and cancellations.

Create a simple matrix where one list is the vertical axis, and one is the horizontal axis, like the example below.

FSPF / Test Script	1	2	3	4	5	6	7
A							
B							
C			X				
D				X			
E					X		
F						X	
G							X

To detect forgotten orphans, process every specification in the upstream asset by finding its counterpart in the downstream asset. If the downstream asset contains a match, place an X in the intersection to show that the specification is not an orphan. If the downstream asset lacks a supporting specification, the matrix exposes an orphan. In the example above, "region" is a forgotten orphan because it appears in the upstream asset (FSPF) but not in the downstream asset (it fits into Test Script B). The implication is that stakeholders expect the customer experience to include "region," but the Test Script excludes it.

To detect gold-plated orphans, process every specification in the downstream asset by finding its counterpart in the upstream asset. If the upstream asset contains a match, place an X in the intersection to show that the specification is not an orphan. If the upstream asset lacks an originating specification, the matrix exposes an orphan. In the example above, "dietary restrictions" is a gold-plated orphan since it appears in the downstream asset (Test Script A) but not in the upstream asset (FSPF). The implication is that the Test Script includes something not previously included in the customer experience.

The default remedy for both kinds of orphans is to treat the upstream asset as truth, preserve its specifications, and revise the downstream asset to match. The other scenario—keeping the specification in the downstream asset—effectively declares that the upstream asset has a defect, and that the project has rework. Even though the rework might require weeks of work, this pre-go-live surprise is preferable over a post-go-live surprise.

As you revise the upstream and downstream assets to remove orphans, update each Traceability Matrix. A Traceability Matrix is done serving its purpose when it contains no orphans. This requires that every row and column of the matrix has at least one X. This scenario is evidence that everything from the upstream asset was remembered and nothing from the downstream asset is gold-plating the project.

Skeptics of Traceability Matrices feel they are overkill, perhaps because they have yet to experience orphans or else are not bothered by them. Skeptics must, instead, pay attention to missed requirements, scope creep, and testing and training defects. Why? Because these scenarios are negative surprises. They are expensive and time-consuming to fix, resulting from unapproved and altogether wasteful work. Traceability Matrices minimize the probability and magnitude of negative surprises by insisting on this two-way accountability.

The ideal timing for the six Traceability Matrices is immediately after the team completes each downstream asset listed above. Teams accustomed to using this asset might even build the matrix concurrently with the downstream asset.

Assignments of Five Verbs for each matrix should combine the assignments of the two involved assets. Both groups of employees should inspect for forgotten items and gold-plating and be involved in the reconciliation. Using the FSPF → Data Model Traceability Matrix as an example, a junior database analyst *drafts* the asset. Experienced data architects *review* and *revise* it, the project sponsor and a cross-company data expert *approve* it, and the junior database analyst *distributes* it. Based on the asset, responsibilities can shift. With the FSPF → Code Traceability Matrix, a junior developer *drafts* the asset, an experienced developer *reviews* and *revises* it, a senior technology employee *approves* it, and the junior developer *distributes* it.

Traceability Matrices shape your culture's discipline because the assets hinder *wasteful* phantom projects, which helps the company's *economics*. Discovering and fixing orphans early keeps *quality* high. The assets shape your culture's empathy. The matrices exercise *bang-the-table* leadership that builds *trust* in the asset portfolio.

Process

The goal isn't just to get there. It is also to have fun along the way.

~ Xavier Dagba (birthdate unknown), Canada-based life coach
and emotional alchemist

I n the twenty-first century, "data"—content—is all the rage. "Content is king" is an excellent cliché for some professions, but *context* is queen, and she indeed steers the customer story and experience. In project work, context translates to *process*. Projects need future state process (FSP) assets.

A typical project team uses the word *process* generously—in conversation. Talking about process is among a team's most natural, educational, and exciting topics. But repeatedly explaining processes is expensive. A team reduces the cost of frequently educating and collaborating about processes by emphasizing the proper documentation of its processes.

A typical business has ambiguity, waste, and dysfunction; that is, something is always broken. When a process is broken, the waste is compounded and expensive. Broken processes repel revenue-generating customers, keep costs high, and hurt stakeholder morale. Broken processes are such a common problem that an entire profession was invented called "Continuous Process Improvement." Nothing else in innovation is such a drain on profitability as broken processes.

Neglect of process documentation is common for a few reasons. It's painstaking work. High-quality process documentation must include exceptional attention to detail. Many innovation professionals enjoy and see value in working "in the weeds," but those who don't render the work unattractive and difficult. Many projects have a culture that skimps on detail, especially if there is no penalty for half-hearted or mediocre work. Transparency of process detail can professionally threaten or embarrass someone, so some stakeholders have the incentive to keep processes ambiguous. Finally, high-quality process documentation always involves many stakeholders. The work is always about someone else, so it requires setting aside personal interests to elevate the interests of others—employees, customers, and more distant stakeholders. A team whose process documentation reflects ruthlessly strong attention to detail overcomes embarrassment while elevating others' interests, and this is uniquely valuable to its customers.

A crucial positive trait of FSP assets is that they are the primary hosts for the project's creative detail. Low-detail creativity exists upstream in the Change Log and Project Charter, while downstream technology assets are constrained in their creativity. FSP detail is instrumental to removing *points of pain* and adding *moments that matter*. Writing the future state resembles writing a play with actors, actions, and a plot. The characters' stories improve with every rewrite—"Continuous Process Improvement." A project team that is passionate about its customers' stories can overcome aversion to precision. Creative detail in FSP assets forms the nucleus of a positive customer experience.

Skeptics of FSP documentation might emphasize data at the expense of process, or their ambition is not on behalf of customers and the customer experience. Ways to win over skeptics include keeping projects' scope small and showing (with the Use Case Assessment) that the impact can still be high. A pattern of small wins, confidence, and improved morale builds momentum and shows skeptics that FSP assets (not data) are the tipping point to making innovation success inevitable.

When a project is done, FSP assets (with some merging) become current state process (CSP) assets. This keeps an organization on its innovation frontier and out of documentation debt. Rigorous process assets minimize negative surprises (for the team and for customers) and contain the seeds of *positive surprises* for customers and stakeholders.

Two more ingredients of the customer experience to consider are pauses and rest. One type of customer interaction is a single, long session. Other customer experiences involve multiple short sessions. All FSP assets should consider and include breaks and rest for customers and stakeholders.

FSP assets shape your culture's discipline by *easing* the customer and employee experiences. The assets reduce *waste* and encourage thoughtful *variability* in the customer experience. The assets shape your culture's empathy by reducing *messiness*, rewriting *stories* for stakeholders, and defining new *moments that matter*.

Customer Lifecycle Model

An Illustration of customers' governed lifecycle states, triggers, and transitions.

Customers have differences; rarely do customers form a homogenous group. They are naturally dynamic and rarely static. Because of this, it's important to segment customers and customize the interactions that each segment has with your business. The Customer Lifecycle Model (CLM) is the single asset to accomplish both.

The CLM combines two frameworks that have been around for decades. A framework called "Object Model" originated in software work, and another framework called "Sequential State Diagram" originated in engineering work. The Elegance methodology repurposes their themes to govern the customer experience lifecycle and, more generically, the stakeholder lifecycle model.

The typical innovation team builds for short-term interactions among customers and stakeholders. The CLM insists innovation teams build instead with customers' long-term interactions—customers' complete lifecycle—in mind. Whereas the typical innovation team emphasizes content for the short term, the CLM emphasizes customer context for a long-term relationship. Short-term thinking often leads to customer behavior that surprises the business. Building for customers' complete lifecycle reduces surprises and eases a company's continual innovation so it can serve customers repeatedly and for a long time. The CLM is the first project-specific asset for every project and imposes customer centricity on every project.

A CLM contains three components: states, triggers, and transitions. For Outdoor Adventures, as an example, customers' states could be *prospect, rookie, experienced,* and *disengaged.* Triggers might include *a first purchase, a third purchase, no purchases for one year,* and *an inquiry.* Figure 10 shows this CLM.

Figure 10. Example of Customer Lifecycle Model.

Formalizing a customer's state facilitates customer segmentation because each state suggests that the company can present unique experiences to the customer.

A CLM also governs which transitions are valid. This example suggests the company recognizes these four transitions, and only these; that is, a transition between "Prospect" and "Experienced" camper is prohibited because it's illogical.

A customer-centric innovation team builds and maintains a CLM for all stakeholders it wants to influence. Customer-centric teams segment stakeholders to whatever degree they customize experiences for those stakeholders.

A young company or innovation team is likely to revise its CLM frequently. A mature company likely changes its model less frequently. In either case, every project warrants a proactive CLM review so that every project emphasizes customer centricity.

Innovators are also never static. Over months and years, an empathetic team expands the asset's scope—adding adjacent stakeholders when the company's indirect impact on them uncovers opportunities for service, value, and revenue. This scenario turns a CLM into an SLM—a *stakeholder lifecycle model*. When you keep the scope strictly to customers and stakeholders, the asset is stable, and maintaining it is low effort and cost.

Skeptics of the CLM accept being ignorant about—and tone-deaf toward—customers. Skeptics shrug how the asset clarifies, segments, serves, and governs customers' evolution. This willful ignorance is laborious and expensive—it not only raises costs but also misses out on revenue opportunities. Thoughtfulness in your CLM exposes revenue opportunities to pursue and costs to avoid.

The CLM's simplicity is another excellent opportunity for a junior employee to *draft* the asset. The best assignments to *review* and *revise* the CLM are for sales, account management, and customer service employees who work closely with customers at the various lifecycle stages. A senior sales leader *approves* the asset, and the junior employee *distributes* it to interested noncontributing stakeholders.

The upstream assets of the CLM are the Customer Experience Hierarchy and the Project Charter. The downstream assets are Future State Scripts.

The CLM shapes your culture's discipline by increasing *variability* when doing so can increase revenue and decreasing *variability* when doing so can reduce costs. Managing variability improves a company's *economics*. The CLM also shapes your culture's empathy. The asset is a form of *listening* to customers and stakeholders, *harmonizing* with their context, and *stewarding* them along a complete relationship lifecycle with the company.

Future State Scripts

> The text-only sequences of the improved customer or stakeholder experience.

Future State Scripts (FSS) detail the actors and actions of the improved customer experience. This asset makes rewriting the customer story as simple as possible.[35] Leaning into the metaphors of the customer story, the art of storytelling, and the performing art of theater, the team can pursue creativity, empathy, and moments that matter. The number of FSS a project needs depends on the Project Charter. The Project Charter specifies the impacted use cases, and every impacted use case requires a script with the same title.

Early in a project, a typical innovation team doesn't think about scripts, storytelling, or theater. A software-oriented team thinks in terms of requirements, business rules, and user stories. These documents are better than nothing, but teams typically form them via stream of consciousness. Those documents emphasize content, technology, and data. In contrast, FSS emphasize actors and context to cultivate empathy for stakeholders. FSS emphasize actions and sequence to cultivate discipline among stakeholders. The simplicity of

---

35 For some innovation teams, drafting a new process flow is "doing too much at once." Breaking the work down into two steps makes the work easier. The first step (FSS) helps a team focus on the language, sequence, and assignments of the action. The next asset (Future State Process Flows) handles cosmetics, that is, sizing and spacing of text and boxes.

FSS undermines process ambiguity and complexity, and process clarity and simplicity encourage automation.

Typical innovation teams also do not build and reap benefits of Current State Scripts. But when FSS benefit from CSS, teams can be precise about the scope and nature of changes to the customer experience—resisting scope creep.

Like CSS, FSS contain two main columns—Actor and Action—which eliminates any concern about the format or cosmetics of the asset. Like CSS, FSS are interim deliverables, making documenting Future State Process Flows (FSPF) easier.

Work on FSS should take advantage of all the benefits of the current state assets. Without those assets, it's more difficult to define scope and know where to stop documenting future state when out-of-scope processes pop up in conversation. Low transparency and unclear boundaries make projects vulnerable to scope creep and becoming risky big projects. Transparency in current state assets gives a team clarity and confidence to keep the scope of FSS modest. FSS include only what they change. The combination of current state assets and FSS combats scope creep, keeps projects small, and keeps risk low. The upfront cost of current state assets enables the low marginal cost of building FSS. Documentation-heavy projects have a reputation for requiring long periods (months) of documenting future state. The small scope in FSS enables this work to be completed in a few days.

As a team drafts FSS, they should reference the Use Case Assessment to improve the particular dimension that received a Red or Yellow rating. Therefore, FSS aim to improve the actor list, the number of steps or handoffs, process duration, effort, or fragility.

The asset resembles the CSS.

#	Actor	Action
1		
2		
3		

Skeptics of FSS tend to dislike different aspects of the asset. Some dislike the script's low ambiguity, simplicity, and ruthless transparency. Some feel it's an unnecessary interim step toward FSPF. Some oppose the script's proposals to increase workload, or decrease workload, believing they can cause job strain or job insecurity. Explain to these skeptics that the asset typically reduces errors and rework for the project team. FSS are a step toward delegating repetitive and undesirable work to technology, and they generally raise the value of employees' post-project routines.

FSS contain exceptional detail reflecting unusual discipline, empathy, and vulnerability. If the project team cannot rewrite the script to meet improvement expectations, the value proposition evaporates, and the team must abort the project. Therefore, FSS help projects "fail fast and fail small" before much work is done.

Because FSS are the first project asset that defines future state detail, executing Five Verbs might have more conflict than current state assets. The documentation aims to foster task conflict and minimize personality conflict. A junior employee can *draft* the asset by formally and informally collecting stakeholder ideas. Stakeholders who *review* and *revise* the asset should have intimate knowledge of what customers and stakeholders want. Involve diverse and the most impacted stakeholders; they are more interested in the detail than anyone. The project sponsor should *approve* the asset and be prepared to serve as tiebreaker among disagreeing stakeholders. The junior employee can *distribute* the script to noncontributing stakeholders.

Upstream assets are CSS, Use Case Assessment, and Customer Lifecycle Model. The downstream assets are FSPF.

FSS shape your culture's discipline in that the asset's *simplicity* keeps the project team's *speed* high. Its attention to detail generates high-*quality* process improvements. Its *traceability* to the value proposition minimizes *waste* by revealing projects unable to fulfill the intended value proposition. FSS shape your culture's empathy. Their transparency builds *trust* and *confidence* in the work. The modest scope *"brings a brick, not a cathedral,"* avoiding the risk of large projects. It *sets the table* for contributors to *co-create* in writing a better *story* for customers and stakeholders.

Future State Process Flows

> A visual form of future operational detail,
> business logic, and process branches.

The next asset to shape future state processes are Future State Process Flows (FSPF). The structure and cosmetics of this format help the innovation team tune the customer experience and optimize assignments, steps, handoffs, and more. The number of process flows that every project needs should match the number of FSS and the number of impacted use cases specified in the Project Charter.

Typical innovation teams execute software-centric methodologies that emphasize content, technology, and data. They don't emphasize the context across human and technology actors. Upstream assets in the Elegance methodology emphasize context and govern the FSPF so tightly that it's difficult for a project team to stray from the value proposition. FSPF shape future business activity with such role transparency that negative surprises (such as assignment overloading or excessive handoffs) become increasingly rare.

Structurally and cosmetically, FSPF resemble Current State Process Flows (CSPF)—swimlanes, actions, decision diamonds, and arrows. FSPF differ from CSPF in the empathy they display and the emotion they can evoke. FSPF show stakeholders more tangible evidence of process improvement and customer centricity. The asset shows the team's creativity and commitment to the customer experience. The visually appealing format isolates improvements and excites invested stakeholders. Indeed, Future State Process Flows are the centerpiece of the Continuous Process Improvement profession.

Four types of FSPF skeptics exist. One type is generically resistant to change. Another type wants to prevent simplification of the process because simplification could jeopardize their job security. A third type objects to the sophistication of the process because that impacts stakeholder responsibilities and interdependence. A fourth type prefers process ambiguity and dislikes that FSPF reduce the ambiguity of accountabilities.

Over multiple projects, customer-centric processes rarely stay static in their sophistication. Teams continually discover reasons and ways to increase or decrease complexity to suit customers' tastes, increase revenue, or cut costs. FSPF are the asset most instrumental in avoiding overcomplicating and oversimplifying processes because both extremes are prone to a team revising them toward an acceptable middle.

FSPF are also the asset essential to automation. Scripts are one-dimensional and can hide certain information. But process flows are two-dimensional and can reveal information. Simple logic and repetitive tasks in a human actor's swimlane can stand out and, at least on paper, seem easy to reassign to a system actor's swimlane (automating it). Conversely, but less common, if a process warrants human touch, judgment, or intervention, with FSPF a team can easily move boxes (reassign tasks) from a system actor to a human actor. Embracing FSPF and automation are signs that a team is trading short-term job security for project success, profitability, and long-term career security.

Two upstream assets encourage FSPF to stay small. Past work on CSPF helps avoid duplicate documentation in FSPF. And because Future State Scripts (FSS) are small, so are Future State Process Flows. These benefits should ease the worries of innovation professionals who feel the high emphasis on documentation leads to massive projects and to obsolete features.

Like FSS, FSPF are subject to conflict among stakeholders, especially because the visual format allows actions and assignments to jump out at the audience more than in script form. Assign Five Verbs to the employees, customers, and stakeholders who have the most at stake in the project. This maximizes the chance that the project will be committed to the highest quality customer experience and minimizes surprises for adversely affected stakeholders. A junior employee can *draft* the asset. Multiple invested, resistant, and enthusiastic stakeholders should *review* and *revise* the asset. The project sponsor *approves* the asset, serving as tiebreaker among disagreeing stakeholders. The junior employee *distributes* the asset to noncontributing stakeholders.

Some conflict (task conflict, not personality conflict) at this stage in the project is a healthy sign. It shows that team members and stakeholders are engaged and take their assignments seriously. Rigor this early in the project reduces negative surprises later. When a surprise does arise, the usual (and most painful) origin of the error is among future state process assets. Errors found later in the project require significant rework to fix and compromise the customer experience when not fixed. In comparison, "errors" (more like disagreements) found early in a project require minimal rework. Early fixes equate to reconciliation, compromise, mutual understanding, and optimizing globally.

In the worst-case scenario, when conflict cannot be resolved, if the project sponsor doesn't play tiebreaker, the project should not proceed. Canceling a project late in its work can devastate economics, morale, and credibility, whereas canceling, suspending, or just pausing a project early might be merely embarrassing. But pausing also shows discipline, awareness, courage, and empathy to stop an entire project team that is set up to fail because of unresolved differences among stakeholders. Canceling a project this early qualifies as "fail fast, fail small."

Upstream assets are CSPF and FSS, and downstream assets are Use Case Definitions.

The FSPF asset shapes your culture's discipline because every process step continually invites team members to eliminate *wasteful* steps and handoffs. Swimlanes invite team members to continually identify ways to *automate* by changing assignments from human to system actors. The asset shapes your culture's empathy in that role *transparency minimizes surprises* for impacted stakeholders. Teams demonstrate *stewardship* when *diverse* stakeholders have *visibility* into the future state. And the visually appealing format helps stakeholders isolate *moments that matter* that the team expects to soon bring to market.

Use Case Definition

> The interrogation of whether Future State Process Flows fulfill process improvements specified in the Use Case Assessment and Project Charter.

The final project-specific process asset—Use Case Definition (UCD)—is an interrogation of the Future State Process Flows (FSPF). A project builds one UCD for every affected use case (as specified in the Project Charter)— which is also one for each Future State Process Flow. Every UCD has three goals: answer questions about how well the process improvements in the FSPF match the Use Case Assessment (UCA); clarify boundaries with other use cases; and explain FSPF's relationships to the Scorecard.

A typical team doesn't interrogate these goals so explicitly so early in the project. Later in the project, a team might casually reinforce the boundaries of process improvements and verify expectations of impacted metrics. The UCD anticipates and formalizes this Q&A immediately after FSPF are approved and before the project team starts work on any technology assets. This interrogation minimizes surprises for the project's sponsors and stakeholders and maximizes the integrity of and confidence in the project's value proposition.

The modularity of customer experiences is valuable in minimizing wasteful activity. The UCD reflects modularity by citing the use case's parent(s), children, predecessors, and successors. To address the UCA rating about actors, the UCD asks the team to list the exact actors (swimlanes). To address the UCA rating about process steps, handoffs, duration, and effort, the UCD asks the team to count the number of process steps and handoffs and estimate time for duration and effort in the FSPF. The UCD asks the team to specify use case inputs and outputs in terms of data and criteria to start and complete the use case. To address the UCA rating for fragility, the UCD asks the team to explain contingency paths such as primary, peripheral, parallel, and error scenarios. If a simple process has a single primary path and no process branches, a fair answer is "not applicable."

The final three questions posed by the UCD relate to the Scorecard. These numbers that the company periodically monitors fall into three categories: numbers that never change, numbers the company can control (formally called *parameters*), and numbers the company cannot control. How a team categorizes these numbers affects process and technology assets.

Treating these categories correctly saves long-term time and effort, and mis-categorizing these numbers is costly. For example, if a project team spends time building processes and technology to control a number it cannot control (weather-related events), the work is wasteful. If a project team ignores building processes and technology for a number it can and *should* control (maximum registrations), preventable blind spots appear. Treating these three types of numbers correctly saves time for project teams short term and for operational teams long term.

Examples of numbers that never change include condominium units in a high-rise building, weeks in a year (52), or counties in the state of Illinois (102). These numbers don't belong in a Scorecard. An innovation team cares about these numbers, but cannot control them, and they don't reflect company performance. The project team specifies these numbers expecting the company shouldn't or won't ever need to change them. This "hard-coding" during the project is a low upfront cost. Being wrong later (post-project) about these values incurs a high cost.

Examples of numbers the company might want to change (parameters) and can control include, for example, the legal age to rent a car, the number of games in a sports season, the number of employees, or the price of a product or service. The project team should not set these numbers but should define processes so Business and Operations employees can set and manipulate these numbers after the project. Building parameters in process and technology assets during a project has an upfront cost, but this cost is justified because, after the project, changing parameter values is low cost, that is, the use case to change values has a short duration. Parameterization avoids this project cost every time the business wants to change its values. Whether a business conceives these numbers via process assets or Scorecard, they belong in the Scorecard as leading metrics because these numbers reflect company decisions and drive business performance.

Numbers the business wants to change but cannot directly control result from business operations, and they change frequently. Examples include the number of customers, duration of the customer experience, and effort in the customer experience. These numbers likely appear in process and technology assets. But the project does not set these values or define processes for Business and Operations employees to manipulate the values. These numbers belong in the Scorecard as lagging metrics. The company cannot control these numbers but cares about them because they reflect business performance.

Categorizing numbers is sometimes difficult. What is a leading metric for one company can be a lagging metric for another company. Even within a single company, categorizing numbers can be fuzzy. To complete the UCD for a single project, the team must take a stance on the relevant numbers. Changing the category of a number should be prioritized like any other item on the Change Log.

A company changes a hard-coded number to a parameter when it wants to remove dependence on Technology employees, eliminate the need for a project to change the values, and instead give control of the metric to a Business or Operations employee. This reduces the duration of the process of changing a metric. The only disadvantage of parameterization is that it increases the risk of human error or sabotage. It's rare to reverse parameterization, but it's valuable to do so when the consequences of a rogue employee misusing or abusing a parameter are dire. Such a project eliminates the parameter and hard-codes the value in the technology, out of reach of Business and Operations employees.

The UCD prompts the project team to decide which metrics to parameterize. Post-project, it's simplistic (low duration and effort) for a Technology employee to change a hard-coded value. However, a change has high friction because it requires another project that is long duration and high in effort. Parameterization makes it quick and easy for a Business or Operations employee to change the value of a metric; it is a short-duration, low-effort task free of the friction of a project team. Parameterization improves company elasticity. Here is a template for the UCD.

Use Case Information	Detail
Use Case Title	
Use Case Description	
Parent Use Case(s)	
Child Use Case(s)	
Predecessor Use Case(s)	
Successor Use Case(s)	
Human Actor(s)	
System Actor(s)	
Inputs (Data)	
Outputs (Data)	
Entry Criteria / Preconditions	
Exit Criteria / Postconditions	
Expected Quantity of Process Steps	
Expected Quantity of Handoffs	
Expected Duration (Measured in Minutes, Hours, etc.)	
Expected Effort (Measured in Minutes, Hours, etc.)	
List of Branches	
Description and Comments for Primary Path	
Description and Comments for Peripheral Path(s)	
Description and Comments for Parallel Path(s)	
Description and Comments for Contingency Path(s)	
Description and Comments for Exception / Error Path(s)	
Relevant Numbers That Never Change (Hard-Coded)	
Relevant Numbers to Control (Leading Metrics)	
Relevant Numbers to Store (Lagging Metrics)	

Skeptics of the UCD feel it's perfectionistic. Many projects don't inter-rogate process improvements so blatantly. A team understandably becomes "heads-down" in the project mechanics and relaxes on the "heads-up" accountability to the value proposition. Every project aims to take what the UCA reports as Red and Yellow and reengineer those use cases (customer experiences) so the UCA can honestly report them as Green. Instead of perfectionism, transparency of the process improvements appeals to highly competitive and ambitious professionals who want to "move the needle." A relaxed, unquestioning team takes its eye off the ball, whereas the UCD aims to exhaust the project team's ideas for customer empathy early in the project. The UCD is not perfectionistic but seeks to prove that the project is on track to fix the problems mentioned in the UCA.

Skeptics are also tolerant about ambiguity surviving into later phases of a project. The UCD minimizes the risk that a project becomes a cliché of "missed requirements." Professionals who typically work on downstream technology assets have strong attention to detail bordering on perfectionism and paranoia. UCD brings that attention to detail upstream. Answering these questions later in the project might feel easier in the short term, but it's ultimately more laborious and expensive. An unanswered question exposes neglect, a gap, and an imminent negative surprise. A healthy project team minimizes ambiguity by answering these questions early in the project and revising the necessary FSP assets to maintain integrity in the project's value proposition.

All contributors to Five Verbs need high familiarity with the UCA and FSPF. With high familiarity, a junior team member can *draft* UCDs. The most impacted stakeholders, including the senior point person of each impacted use case, are assigned to *review* and *revise* UCDs. The project sponsor *approves* the UCDs. A junior employee can *distribute* the script to noncontributing stakeholders.

Upstream assets are the UCA, Project Charter, and FSPF. Downstream assets are high-level design.

The UCD asset shapes your culture's discipline by enforcing *vigilance* about the value proposition and formalizing the *autonomy* of every use case. The asset shapes your culture's empathy by drawing *boundaries* between customer experiences, maintaining *self-awareness* among business operations, and *banging the table* so that the project is on track for its value proposition.

People

I consider my ability to arouse enthusiasm among men the greatest gift
I possess. The way to develop the best that is in a man
is by appreciation and encouragement.

~ Charles Schwab (b. 1937), US investor and financial executive

One cliché about troubled projects is that the team undercommuni-
cated. People inside or outside the team might have neglected to
communicate. This undercommunication cliché stares every project team
in the face. "People" assets are critical to prevent undercommunication and
avoid reinforcing the cliché.

People assets support the profession and formal discipline called "Change
Management." A typical project team is aware of Change Management.
But even knowledgeable teams undercommunicate in some ways and
overcommunicate in others, creating noise, negative surprises, and rework.
Worse, some project teams completely neglect Change Management, seeing
the function as a nice-to-have, and disregard it as an easy way to cut costs.
These teams embody the cliché of undercommunication.

The Change Management community educates on several frameworks.
The Elegance methodology piggybacks on a framework called ADKAR
because, compared to other frameworks, ADKAR is most compatible with

Five Verbs and avoids under- and overcommunication.[36] ADKAR stands for awareness, desire, knowledge, ability, and reinforcement. The model proposes that disciplined, empathetic project teams shepherd their stakeholders through these five steps to achieve healthy change adoption. The Elegance methodology proposes one asset for each step of the ADKAR framework, except for the knowledge step, where it proposes two.

People assets are not the only asset group that prevents teams from falling into the undercommunication cliché. Project management and process assets are also critical, but their primary audience is the project team. The primary audience of people assets is stakeholders outside the core project team. These stakeholders are typically on the receiving end of the fifth and final verb, *distribute*. Every project needs all three asset groups to communicate adequately with all stakeholders.

Stakeholder groups—and membership in them—are ambiguous and fluid. It's futile and counterproductive to overmanage them. One person might weave among the core project team, extended stakeholder population, and customers. People assets navigate this ambiguity by involving distant stakeholders who are not exposed to the more detailed process and technology assets. That way, stakeholders with ambiguous interests in a project can stay informed at the level of rigor of ADKAR: they are aware of the project, have a desire for the project to succeed, are knowledgeable about how the change impacts them, have the ability to adopt the new processes and technology, and can reinforce the changes.

Skeptics of people assets don't mind surprising, excluding, or undercommunicating to stakeholders. Skeptics don't see the damage or feel any consequences of communication neglect. In pursuit of cutting short-term costs, they don't see that people assets do more than mitigate the downside effects of poor communication. People assets have significant upside that other assets don't. Whereas other assets must be applied with rigor and caution,

---

36 ADKAR was created by and is the intellectual property of a company called PROSCI. Jeff Hiatt is the author of the book *ADKAR: A Model for Change in Business, Government, and Our Community (2006)*.

people assets generate trust, enthusiasm, and goodwill. People assets are not solely for avoiding negative sentiment and surprises, though; their goals include increasing adoption and revenue.

The project manager and the change manager are essential roles in every project team. The boundaries of these roles vary across people, projects, and organizations. Like any two roles, these roles risk working in silos. People assets are a foundation for navigating differences, setting boundaries, and guiding collaboration between the roles.

People assets shape your culture's discipline because they *ease* engagement and adoption for stakeholders. They accommodate the *elasticity* of stakeholders who meander near and far during a project. The assets shape your culture's empathy. People assets tell audiences that improved *experiences* and *performance* are coming. The assets *include* the audience in communication along the way and *celebrate* a project's success when the project is completed.

Awareness Blast

> A memo to team members, stakeholders, and prospective stakeholders announcing the start of a project and minimizing undercommunication about it.

The first step in the ADKAR method is awareness, and the first "people" asset strives to make stakeholders aware of the project. Because the number of recipients is often high and the message is often short, the asset is titled Awareness Blast.

A typical project team holds a project kickoff meeting that covers positive expectations of the project. A typical team casually circulates emails about a project starting. But a typical team doesn't craft an email that explicitly tries to reach interested and impacted stakeholders. The Awareness Blast is an early step to avoid a project falling prey to the cliché of undercommunication. The asset aims for inclusivity and engagement.

The Awareness Blast should not duplicate the Project Charter, which aims to be exhaustive in content. In contrast, the Awareness Blast aims to be exhaustive with inclusivity and engagement. It has an inviting and promotional tone. Below is a starting point for an Awareness Blast.

Hello! You are receiving this memo to inform you about a project that is just starting. The project's title is ABC. Its primary objectives are XYZ. Its scope is limited—we're expecting to include and impact PQ, but not RS.

Origins of the project include circumstances DEF and events JKL. Given other innovation priorities, now is the right time to start this project.

A core project team is forming, and the team believes you should know about this project. You are a potential stakeholder and may know other stakeholders who should know about the project. The core team would like to know how involved you wish to be in this work. ZZZ currently manages the project's stakeholder list. Please contact them to discuss the nature of your involvement with the project and your recommendations to involve other stakeholders.

The company is excited to start this innovation. The core team welcomes and appreciates your support, engagement, and contributions along the way.

(Signed) Project Sponsor

Skeptics of the Awareness Blast have fair concerns, including those about confidentiality and sensitivity. There are such things as too much transparency and too many cooks in the kitchen. Need-to-know and right-to-know policies matter, and the body of the Awareness Blast should clarify that its purpose is to engage legitimate stakeholders rather than spread universal public knowledge about the project.

But some skeptics are generic "communication minimalists" who increase the project's risk of undercommunication. Some skeptics overestimate the risk of overcommunication with an Awareness Blast and underestimate the high cost of undercommunication. They underestimate the upside of proactive and vulnerable communication that improves a project's stakeholder list and Project Plan assignments. An Awareness Blast is a simple, low-cost asset that aims for high engagement and low risk.

Healthy teams execute Five Verbs for the Awareness Blast. A junior employee can *draft* the asset, and core team members *review* and *revise* it. The project sponsor includes their signature in the memo and *approves* it. The junior employee *distributes* or "blasts" it to the project's extended stakeholder list (defined in the Project Charter).

Your Awareness Blast is successful when stakeholders respond to it. Responses to the blast likely lead to revising the Project Charter—especially the *stakeholder list*—and revising *assignments* in the Project Plan. And later in the project, all stakeholders keep project awareness, inclusivity, and engagement high.

The Awareness Blast does not have a simple, single upstream asset. Instead, start *drafting* it as you are completing a *draft* of the Project Charter, and when the charter has enough content to help draft the Awareness Blast. For example, the stakeholder list in the Project Charter identifies the audience for the Awareness Blast. Soon after you distribute the Awareness Blast, the Project Charter will stabilize and be ready for approval. The downstream asset of the Awareness Blast is the Elevator Pitch.

The Awareness Blast shapes your culture's discipline. Early stakeholder engagement enables a project to avoid silos, negative surprises, and negative attitudes. Optimizing across diverse stakeholders improves project *quality*. The asset shapes your culture's empathy because its audience and message show *positivity, inclusivity,* and *stewardship* beyond the core team, for extended stakeholders.

Elevator Pitch

> A short statement that touts the value proposition
> and generates enthusiasm for the project.

The second step of ADKAR is desire. The corresponding asset, an Elevator Pitch, aims to increase stakeholders' enthusiasm for the project mission and desire for project success. A project's Elevator Pitch emulates a personal Elevator Pitch—a short explanation of why a stakeholder would support and engage in the business relationship.

A typical project team informally has an Elevator Pitch. The team likely has many versions that it uses inconsistently and that even change during the project. A single, thoughtful Elevator Pitch is a great tool to set expectations and manage perceptions without getting into the details of the Project Charter, process flows, or training materials. The asset is an inexpensive way to build engagement, generate enthusiasm, and repeatedly win over project skeptics and those resistant to change.

Stakeholders assume practically every project is profit-minded, so an Elevator Pitch should avoid referencing the company's or its executives' financial welfare. Instead, every Elevator Pitch should explain a project's benefit to customers or employees. Consider these topics: health, security, convenience, prosperity, community, family, fun, adventure, and memories. The topics resemble those of Maslow's Hierarchy of Needs and refer to ordinary human experiences of suffering and delight.

Because an elevator ride is rarely very long, an Elevator Pitch should be shareable in less than twenty seconds. Structurally, this limits an Elevator Pitch to around three sentences. In an Elevator Pitch, a project's what is a nice-to-have, and a project's why is a must-have. You could structure an Elevator Pitch to resemble an asset explained earlier in this book—the ORS Report, which encompasses observations, reactions, and suggestions. Here are a few examples of great Elevator Pitches:

> Automobile fatalities have been steady for a decade, and the public feels they are too high. Drivers have shown acceptance of additional safety features. Starting in 1955, all vehicles will be manufactured with seat belts, which have been proven to reduce fatalities.

> With the decline of company pensions and all the financial security they bring, our company continues to improve investment opportunities for workers. This month, we introduce ten new mutual fund options and more flexible contribution plans to save and invest for retirement.

> Baseball fans used to face tough choices between long lines to buy food (and risk missing a big play on the field) and settling for limited food options from strolling vendors. Starting this season, fans can order any item from an app on their smartphones and have it delivered to their seats. They'll never miss a play *and* never miss out on their favorite ballpark snack.

Skeptics of an Elevator Pitch perceive it as cheesy, altruistic, even idealistic. But the exact point of an Elevator Pitch is marketing, advertising, and promotion. Skeptics must remember that many projects have stakeholders who hope the project stops or fails. This asset acknowledges that some stakeholders might disapprove of the project. Many skeptics see change as a threat and prefer the status quo. The Elevator Pitch is an efficient, edgy tool used to reduce the perception that the project is a threat or unimportant. The asset helps every project compete with the status quo.

A team applies Five Verbs as for any other asset, with a junior employee *drafting* it. What is unique to the Elevator Pitch is the value of having stakeholders with *high* enthusiasm, *low* enthusiasm, and even skepticism *review* and *revise* the Elevator Pitch. The diversity of those stakeholders reduces overpromising and improves the integrity of the Elevator Pitch's claims. The project sponsor *approves* the Elevator Pitch, especially because they serve as a point of escalation for stakeholders and a tiebreaker between diverging ideas. The junior employee can formally *distribute* the written version, and, of course, every stakeholder should informally deliver the pitch when the right occasions arise.

Upstream assets of the Elevator Pitch are the Awareness Blast and Project Charter. Its downstream asset is the Training Approach. The Elevator Pitch and future state process (FSP) assets inspire each other, so a team might complete them in parallel. The Elevator Pitch provides a North Star as a team drafts FSP assets. Later, when FSP assets capture stakeholder moments that matter, the team might revisit and revise the Elevator Pitch.

The Elevator Pitch shapes your culture's discipline. It promotes project *quality* by emphasizing customer centricity and the value proposition. It reduces the *variability* of the project's real and perceived goals. The asset shapes your culture's empathy. It shows that the team is passionate about customers' *stories* and wants to create customer *moments that matter*. For a project team that wants a positive *legacy*, the Elevator Pitch is its bull's-eye.

Training Approach

The expectations for training
materials and delivery.

The letter *K* of ADKAR stands for knowledge. Transferring knowledge about the imminent change equates to training impacted stakeholders. It makes no sense to standardize training materials for every project (so the asset doesn't have its own section in this book), but it makes sense to standardize certain training decisions that every project must make. The asset for standardizing these decisions is a Training Approach.

A typical organization standardizes some training activities. For example, scheduling, registration, and venue often feel routine. But a formal Training Approach exhausts the questions every project must answer to minimize neglect and surprise for trainers and trainees. Some answers vary for every project, and some answers stay the same. The Training Approach shows a history of answers for multiple projects to ease decisions for every subsequent project.

The Training Approach asset contains information for multiple projects, so, first, capture the project title. For every project, describe the training audience and estimate its size. A complex project might have more than a single homogenous audience, so consider whether different audiences need different answers for training decisions. Describe the format of the training materials, that is, the combination of video, text on a screen, and printed materials. Assign a training administrator to maintain the training materials and serve as a first point of contact for questions.

Include names and the sequence of speakers in the training sessions. If you expect to record the training session, assign who is recording and posting those files. Share the repository location so stakeholders can access the recording. Share the date range during which the recordings are expected to be available. If a physical location is needed, identify the venue such as by giving a conference room name or number. If training is to be delivered virtually, specify the software to be used.

Some projects can deliver training in a single session, whereas others might conduct multiple sessions, partitioning audiences or content. Estimate the calendar window of the training sessions (e.g., sessions within ten business days) and the duration of each session (e.g., all sixty-minute sessions).

Some training sessions are scheduled a stated number of days away from a project milestone, such as a testing completion date or static go-live event. The Training Approach should record this "scheduling anchor." Training sessions typically involve invitations, so assign who manages them and whether they monitor attendance.

And, finally, include whether the training materials contain questions or tests, which might be at the start or the end of the training session or distributed within. Some training sessions are interactive enough to justify training data or a training environment—if so, the asset should reflect this.

All this information fits into the template below.

Specification	Project 1	Project 2	Project 3
Title			
Number of Audiences			
Audience Description(s)			
Audience Size			
Format of Training Materials (Video, Text on Screen, Printed Materials)			
Administrator of Materials			
Speaker List and Sequence			
Administrator of Recording and Posting			
Repository for Recording			
Estimated Dates for Recording Availability			
Location of Sessions (Room Number or Software for Virtual Session)			
Administer of Room Reservations or Virtual Session			
Quantity of Training Sessions			
Distinguishing Training Sessions (Audience or Content)			
Scheduling Anchor			
Calendar Window			
Duration of Each Session			
Administrator of Session Invitations and RSVPs			
Monitor Attendance? (Y/N)			
Administrator of Sessions Attendance			
Tests or Questions in Materials? (at the Start, End, Distributed)			
Training Environment or Training Data?			

Skeptics of a Training Approach have not been exposed to the surprises that arise when a project neglects these questions—or the surprises don't bother them. The Training Approach asset is another example of the reality that ignoring questions doesn't make the work disappear. Ignoring questions only perpetuates ambiguity and postpones alignment on the decisions. For every project, the questions are the same, and many of the answers are the same. The transparency of the information in the Training Approach asset eases alignment on these answers.

A team applies Five Verbs as for any other asset, with a junior employee *drafting* the column for their team's single project. Team members who *review* and *revise* the column for their project know the names and descriptions of the stakeholders, the training audience(s), and what will be effective for each training audience. The project sponsor *approves* the project-specific information. The junior employee *distributes* the information to trainees and other noncontributing stakeholders.

A cross-column inspection is worth monitoring for consistency of training work across projects. Many companies have a Learning and Development (L&D) department. A junior L&D analyst can *review* and *revise* the asset to reflect when a project has unique training needs. A senior L&D employee *approves* the asset within each project—inspecting across projects. The junior L&D analyst can *distribute* the asset to all L&D employees.

The upstream asset of the Training Approach is the Elevator Pitch. Especially for sensitive and controversial projects, the Elevator Pitch often influences the hands-on, hands-off, or "intimacy" of training sessions. The downstream assets are the training materials, which serve as the other asset representing the *K* in the ADKAR model. The Elegance methodology doesn't emphasize training materials because standardization isn't applicable here, but it does acknowledge them as an asset. Teams rarely neglect training materials, and skeptics of them are rare. That said, asset sequence still matters. Training materials have several upstream assets such as the Project Charter, Customer Lifecycle Model, Future State Process Flows, User Interface, Report Inventory, Report Detail, and Training Approach.

The downstream asset of training materials is the Go-Live Announcement.

The Training Approach shapes your culture's discipline. Standardized questions and transparency to other projects maximize *speed, quality,* and *ease* and reduce the *variability* of these training decisions. The asset shapes your culture's empathy because information about upcoming training improves *trust* in the project team and the training materials. The Training Approach minimizes *reinventing the wheel* for this step of ADKAR.

Go-Live Announcement

> A memo (usually celebratory in tone) to team members and stakeholders announcing that the new processes, technology, and customer experience are in effect.

The next step in the ADKAR framework is ability—the ability to adopt the new customer experience, execute new processes, and use new technology. These abilities require a "go" signal from the innovation team, and the corresponding asset is called a Go-Live Announcement.

Go-live events can be terrifying, but typically they signal tremendous accomplishments of innovation teams. Go-live is an occasion to celebrate.

A typical project team issues a Go-Live Announcement, but announcements vary across projects, teams, and companies. One way they differ is in the level of detail. A detailed announcement includes accomplishments, appreciation, relevant dates, and contact information for handling questions and feedback. It might resemble a less-structured Workstream Status Report. The announcement might anticipate questions and explain nuances or exceptions in the new way of working.[37] The asset should set expectations and manage perceptions of the new way of working as transparently and vulnerably as the team sees fit.

---

37 Time-sensitive questions and exceptions belong in the Go-Live Announcement. Timeless questions and exceptions belong in FSPF and training materials (not an FAQ, as explained in Appendix B).

Announcements also vary in tone. An optimistic announcement conveys confidence and enthusiasm in the new customer benefits, experience, processes, and technology. It gives a sense of celebration and success. On the other hand, some announcements warrant a humble tone if stakeholders feel negativity about the project, the team, or the company. An announcement warrants a cautious tone if the project helps stakeholders emerge from a crisis. In other words, the asset should avoid being tone deaf. A great Go-Live Announcement exercises sentiment awareness for the project.

Below is an example of a detailed, subdued Go-Live Announcement:

> This message is intended for stakeholders monitoring Project ABC whose goal has been to improve situation XYZ. Over the past few weeks, the project team has completed the build, test, and training for the relevant processes and technology. Our PQR customer segment will see and experience the improvements on Date X. Our TUV customer segment is tentatively scheduled to go-live on Date Y.
>
> As the project sponsor, I've done my best to be aware of and empathetic regarding the problems the team identified that I believe are now improved. I appreciate the contributions, collaboration, and perseverance of everyone involved. If you have questions or feedback, please reach out via email.

Here is a generic, dry, bare-bones starting point for an upbeat Go-Live Announcement. Every project team should rewrite using their personality, enthusiasm, and flair. Be as detailed and prescriptive as you see fit.

> Effective Date X, as users/customers/stakeholders of Experience ABC, improvements will be available to you!

Credit and congratulations to every project team member who contributed and collaborated over the past few months. Thank you for your hard work and dedication. Please send questions and feedback via email.

Skeptics of Go-Live Announcements don't want to overshare, especially any information less than favorable. This hesitation is understandable, but a project sponsor's self-awareness, humility, and transparency spreads credibility and goodwill among stakeholders.

The Go-Live Announcement warrants unique assignments of Five Verbs. The project sponsor *drafts* the asset. For a short announcement for a simple project, the sponsor consults only a handful of core project team members to *review* and *revise* the announcement. For a lengthy announcement or a complex project, the sponsor consults a broader group, such as contributors to the Project Charter, future state process assets, training, and testing. For the most sensitive projects, the sponsor consults the most prominent skeptics and riskiest stakeholders. The sponsor *approves* and *distributes* the Go-Live Announcement.

Upstream assets include training materials, Migration Script, Data Conversion Script, Actual Results, and the Deployment Plan. When it is time to *distribute* the Go-Live Announcement, testing is finished, training is finished, and code and data are in place. The Go-Live Announcement's project-specific downstream asset is the Closure Report. Its project-independent downstream assets are the Voice of the Customer and Seller, ORS Report, Scorecard, I Like I Wish I Hope I Wonder, Change Log, and Parking Lot.

The Go-Live Announcement shapes your culture's discipline. The asset shows the project team's *accountability* because it completed what it started, *alignment* on the outcomes, and *momentum* of a project nearing completion. The asset shapes your culture's empathy by *integrating* many innovation skills to improve customer-centric *performance*. The asset conveys to the project team that their work is worth *celebrating* and calling *GETMO*.

Closure Report

A memo to team members and stakeholders announcing the project is complete and the project team is disbanding; it includes how well the business is realizing the expected value proposition.

The final step of the ADKAR framework is reinforcement. This step intends for the project team to reinforce to impacted stakeholders the importance of adherence to new processes and adoption of new technology. The Elegance methodology sees an opportunity to reinforce the project's value proposition. Lens of the individual assets govern adoption by individual employees. The Closure Report asset educates stakeholders on achieving and monitoring value as the project declares closure.

A typical project team disbands after the go-live event, and members move on to other projects. The typical team does not proactively track the value proposition or generate a Closure Report. This abdicates accountability to the value proposition. When project team members know the Closure Report holds them accountable, they work harder to keep the quality and integrity of the upstream assets high.

Publishing a Closure Report has four prerequisites that easily require one or two weeks to complete. The first task is to merge content from project-specific assets into the appropriate current state assets. Review, revise, approve, and distribute the Current State Inventory, Current State Scripts, Current State Process Flows, System Actor Inventory, Customer Experience Hierarchy, and Use Case Assessment (UCA). Next, for the UCA, note whether ratings improved from Red or Yellow to Green for the use cases impacted by the project. Revise the Scorecard on its usual schedule, noting whether metrics are trending toward the project's goals. The fourth task is to review the Project Charter to see whether the project diverged significantly from any expectations set in the charter.

The project sponsor drafts the Closure Report. Below is a generic, dry starting point. Every project team should rewrite it with an appropriate tone and project-specific detail (such as listing the processes, use cases, and customer experiences impacted by the project).

Dear Stakeholders:

Project ABC went live on Date X with this Elevator Pitch explaining what we set out to do:

<Insert Elevator Pitch>

Adopting new processes and technology was critical to achieving the Elevator Pitch and value proposition. Managers continue to monitor adoption through status reports and other assets. Thank you for your continued cooperation.

Since the go-live event, the project team has completed the following to bring the project to a close. What was the future state a few weeks ago is now the current state, so the team merged project-specific content into current state assets. This gives employees and future project teams a single place to reference to learn about integrated new and existing processes. Second, the team repeated a review/revise/approve cycle of the Use Case Assessment to see how well the project upgraded relevant aspects of impacted use cases from Red and Yellow to Green.

The team concluded that aspects ABC of use cases XYZ improved to Green and aspects JKL remain Red or Yellow. The remaining problematic use cases join existing innovation candidates to consider, prioritize, and scope in future projects.

The team monitors the Scorecard and sees that the "metrics that matter" for this project are (are not) trending as intended. The team also reviewed the Project Charter to reflect whether the project diverged significantly from any expectations set in it. The newsworthy divergence is XYZ.

I have two requests for you as the team disbands. Please continue to monitor project-related metrics in the Scorecard. Also, as you identify additional innovation ideas (related to this project or not), please add them to your ORS Report, the Change Log, or the Parking Lot.

Signed,

Project Sponsor

Skeptics of a Closure Report not only prefer to move on to other projects but also dislike transparency of any shortcomings related to the Project Charter, Elevator Pitch, UCA, or Scorecard. Skeptics also shrug at the rigor and gravity of the prerequisite tasks and how they enforce integrity of the value proposition and set up future projects for success. The Closure Report is evidence that your team avoids documentation debt.

The project sponsor *drafts* the Closure Report. The employees who *review* and *revise* it include team members who contributed to the four tasks related to the report. The project sponsor then *approves* and *distributes* the report.

Upstream assets include the Go-Live Announcement and Scorecard. Downstream assets include the Voice of the Customer and Seller, ORS Report, Change Log, and Parking Lot. Child assets of the Closure Report are the current state assets.

The Closure Report shapes your culture's discipline by maintaining *vigilance* to the project's value proposition. It is evidence of *momentum* in one project team and represents *autonomy* between one project team and the next. The asset shapes your culture's empathy by *bringing out the best* in every project and establishing a *legacy* for every project team. The report imposes *humility* as the sponsor acknowledges that the company has more improvements to pursue. The report fosters *self-sufficiency* and minimizes *messiness* for the next project team.

TECHNOLOGY

You're either the one that creates the automation or you're getting automated.

~ Tom Preston-Werner (b. 1979), American software developer and entrepreneur

Technology takes the eight Ds out of the hands of humans; because of technology, humans can reduce their dull, dirty, dangerous, difficult, demanding, demeaning, delicate, and dear work. Technology improves the quality of life for customers and employees.

Typical innovation teams build several assets comprising design, build, and test work, but only some teams aim for thoroughness or standardization. The typical team has a minimalist approach, that is, do the minimum amount of work that puts new technology in place for customers. Cutting corners on technology assets reduces upfront costs but raises ongoing customer complaints and unglamorous rework for the project team. No business case broadcasts that the team plans to cut corners, but such neglect certainly jeopardizes value propositions of projects.

Many aspects of technology work are glamorous: designing, building, testing, and putting technology into the market. But a minimalist approach rarely produces glamorous results. Minimalist approaches produce thoughtless results. Yet thoughtful teams don't have to wallow in technology assets. The most impressive technologies avoid minimalism of thought. Teams must embrace and immerse themselves in the right assets. The best technology isn't easy. It has a high upfront cost, but it makes work look and feel easy (low marginal cost). Because of the technology industry's demands for customer centricity, intelligence, and perseverance, it is a respected and lucrative profession, one that's clean, safe, and rarely dull.

The primary goal of these chapters about technology assets is not to improve how software-centric methodologies define them; the goal is to educate innovation professionals who are unfamiliar with the assets. Professionals must include these assets in assignments and scheduling. Regardless of their role on a project team, all "innovation-literate" professionals expect their teams to build these assets.

Building a technology asset for the first time might be a big job, but you reduce the risk for every project if you start small and keep every project's scope modest. An underappreciated trait of technology assets is that the discipline in every project keeps the laboriousness of subsequent projects low. Consistent with the spirit and tactics of the Elegance methodology, upfront generosity of time and talent leads to ongoing low marginal cost.

Skeptics of technology assets need more education about the attention to detail that technology needs to work as expected. Humans must tell code what to do about *everything*, for example, which circumstances to automatically turn on and off, how to handle an unauthorized or nefarious user, what to do and tell the user when the device overheats, and countless other scenarios for code to imitate human judgment.

The following sections on technology assets contain less detail than previous sections. After decades of pervasive software-centric methodology, high-quality information, examples, and templates are abundant and publicly available. As a result of decades of usage, more innovation professionals are familiar with these assets—typical companies use them, and skeptics are rare. Technology platforms, tools, and languages change, but these assets are stable.

In the twenty-first century, technology assets shape workplace culture less than the assets discussed in previous sections do. Technology assets are common and commoditized, and they are the last group of assets in every project, whereas the previously discussed assets are less common. They are foundational and distinctive for every project. Because of technology assets' relatively low impact on culture, these chapters rarely mention culture.

Problems originating in technology assets are less crippling to a team and to a company. Consider that problems originating in a Roadmap, Project Charter, or process flow measure rework or lost time in weeks or months, but problems originating in the design, build, or test phases of technology assets measure rework or lost time in hours or days.

Every project must be ready to build new technology assets and revise existing ones last touched by a previous project. Every project must treat every technology asset as a relevant input. Some projects impact every technology asset, and projects that don't must approve decisions about not revising certain technology assets.

Technology assets are exacting and unforgiving. Legacy and the Elegance methodologies are formulated to detect poor precision. Ambiguity leads to immediate workarounds and delays.

Software-centric methodologies promote their culture trait of being iterative. Rhythm, experimentation, and forgiveness have their place, but a culture trait of conscious iteration dismisses the value of precision and tolerates sloppy work. Getting things right the first time reduces iterative rework; it also reduces employee frustration and frees employees to work on more valuable innovations.

Although not new or distinctive, technology assets—organized into four groups: high-level design, detailed design, build, and test—still broadly influence team culture. These assets shape a culture of discipline with their *precision* and *attention to detail*. They shape a culture of empathy by reducing the *laboriousness* of maintaining and improving technology from project to project.

High-Level Design

The first group of technology assets is design assets. A robust innovation methodology has enough design assets to split them into high-level design (HLD) and detailed design (DD). What qualifies as HLD is mundane, that is, its downstream asset is also a design asset.

A typical project team might treat HLD assets casually. The primary reasons are the low penalty for short-term neglect and impatience to get to DD assets. But neglect translates to duplication, disuse, and waste of hardware and software. Neglect in HLD is a version of tech debt upstream from code. Short-term debt is not crippling, but significant debt has consequences, including slowing teamwork, speed to market, and customer experiences. Minimizing and avoiding tech debt among HLD assets is not complicated; reversing a decision is quick and easy.

Another risk among HLD assets is ownership. The assets show which technology actors might have broad usage among processes and employees. Assignments can be challenging, especially for technology actors that haven't changed in a long time. Just as a senior and a junior point person are designated for the Use Case Assessment, every HLD asset should be assigned its own senior and junior point people.

HLD assets shape your culture's discipline by minimizing *waste* (technology sprawl) and keeping *speed* (customer experience and employee experience) high. The assets shape your culture's empathy by minimizing *ambiguity* in a project team's decisions about technology actors.

Technology Architecture

> A diagram to show the relationship among technology actors—hardware and software.

Most technology assets emphasize software. The Technology Architecture (TA) asset is an exception. It emphasizes hardware—the electronics that host the software.

The typical project team deals with the TA asset infrequently. It's managed by someone outside the team or even outside the company. But project teams should have basic competence in what it contains: inventories of hardware devices and their relationships depicted as the direction of information flow, relationships each hardware device has with technology actors outside the company, and software that resides on each hardware device.

As projects progress, code and data are moved to different places called "environments." New, immature, or the least-tested code resides in an environment called "Development," "Dev," or "Sandbox." Mature code resides in an environment called "Test." Approved code supporting actual customer-facing business processes resides in an environment called "Production." The TA asset shows the hardware devices that host these environments. Each environment typically occupies its own hardware device.

The final way the TA asset partitions technology actors is according to the code's proximity to technology users. Code closest to the user is called the presentation layer. Middle-layer code typically contains business logic, and the most distant code layer is typically related to data storage and retrieval. These layers commonly reside on different devices in the TA.

With the rise of SaaS (Software as a Service) and cloud architectures, many companies delegate a portion of their architecture to an outside company. A rigorous TA asset reflects this, and the arrangement might simplify the asset. On the other hand, corporate mergers and partnerships typically connect architectures, complicating the asset.

Neglect of the TA asset results in duplicate or unused hardware. The waste in both scenarios increases costs of device monitoring, electricity, and storage.

In building the TA asset, assign an experienced technical architect to *draft* it. That architect facilitates other architects—junior and experienced—to *review* and *revise* it. A senior technology employee *approves* the asset. The authoring architect *distributes* the asset to the company's contributing and noncontributing technology architects.

The upstream assets are Future State Process Flows and storage and traffic capacity reports from outside the project (and outside the scope of this book). The downstream assets are the Data Flow Diagram and the Code Module Inventory.

Storyboards

The depictions of the customer's visual and emotional experience, governed and progressing through their story toward a goal.

Storyboard assets originated in the film industry but are now used in book writing, advertising campaigns, and innovation. Upstream assets govern process, downstream assets govern data, and in between, Storyboards govern visual experience and customer emotion.

A typical innovation team does not manage customer emotion so proactively or explicitly as this asset does. Conversations might reference customer empathy or sentiment, but typical assets govern process and data. Yet shaping and tracking emotion helps isolate customer highs and lows (i.e., "moments that matter" and "points of pain") and helps determine what to fix, what to protect, and what to emphasize.

At their most dry and pragmatic, Storyboards resemble assets called wireframes. Storyboards contain thumbnail (low-fidelity) visual bundles the user will see and are one step away from software detail that structures a webpage. The asset partitions the Customer Lifecycle Model and Future State Process Flows and organizes the segments into webpage titles (not the detailed webpage design).

At their most empathetic, Storyboards shape the intended emotional journey of the customer. Storyboards engineer fluctuations in mood and depict feelings like intensity, relaxation, buildup of tension, release, and closure. Storyboards contain a plot, explain a narrative, and illustrate scenes of a customer's story as they pursue and progress toward goals. Storyboards aim to systematize customer delight and design positive customer surprises.

Storyboards contain as few as three ingredients: the scenes or webpages, the titles of each, and arrows to show context and relationships among scenes frame by frame. Storyboards are neither text-heavy nor do they require impressive artwork. Stick figures and emojis are adequate. Storyboards minimize visual density to avoid overwhelming or boring a stakeholder in a review.

Neglecting Storyboards misses opportunities to reduce customer points of pain and to capitalize on customer exhilaration. Storyboards are another step in which an innovation team can be attentive and empathetic to customer needs.

Assets like Storyboards relate to the profession of user experience (UX). A junior UX professional can *draft* Storyboards. At least one experienced UX professional and the associated junior point person reflected in the UCA *review* and *revise* them. The project sponsor and the associated senior point person reflected in the UCA *approve* the Storyboards. The junior UX employee *distributes* the assets to interested noncontributing stakeholders. Storyboards' upstream assets are Use Case Definitions and the Elevator Pitch. Downstream assets are User Interfaces.

Data Flow Diagram

> Visual representation of information
> flows among technology actors.

A Data Flow Diagram (DFD) is a visual representation of the information flows among technology actors. Most assets reflect the lens of a human actor or the customer. Instead, the DFD reflects the perspective of the technology actors (or the data itself) as information travels among them (i.e., files, processes, databases, and external entities).

A typical project team that hosts its own technology distinguishes between a logical and a physical DFD. A logical DFD contains a conceptual flow of information describing the *nature* of data and transformations taking place among technology actors. A physical DFD reflects the actual *names* of technology actors (i.e., file names, process names, and database names).

A team often builds DFDs at different altitudes to focus on education and troubleshooting. High-level DFDs might include human actors, but detailed DFDs de-emphasize them and often exclude them altogether.

A DFD is more detailed than a Technology Architecture (TA) asset. The TA asset emphasizes hardware. Every hardware device might host more than one technology actor, and a single prominent technology actor might logically reside on multiple hardware devices. But DFDs have limitations on their detail. They exclude presentation-layer detail, business logic, database calls, and anything related to process steps or sequence.

DFDs help build, configure, and monitor appropriate numbers of files, processes, and databases to optimize the performance of data moving among hardware and software actors. The primary measure of performance is speed. Continually, across a company's Production, Test, and Development environments, network operators monitor the speed and volume of network traffic. When operators detect high volume and slow speed, the DFD is their map to isolate where the bottlenecks exist. DFDs help identify the hardware and software to add and configure to ease traffic flow and keep processing speeds high. When operators are ready to change and configure technology actors, a project revisits and revises the DFD.

Neglect of a DFD asset hinders the speed and confidence in trouble-shooting performance problems. Disadvantaged troubleshooting and fixing hurts the customer experience. Neglect erodes the value of hardware and software with redundant and obsolete files, processes, and databases. This leads to unnecessary or gridlocked network traffic, shadow processes, and poor system responsiveness. Neglect of the DFD is a data governance problem that causes process governance problems.

Assets like the DFD need employees with titles such as "data architect" and "database administrator" (DBA) to administer them. A junior DBA can *draft* the DFD. Team members who *review* and *revise* the DFD include at least one experienced architect. A data expert with a broad understanding of the company's technology actors *approves* the DFD. The junior DBA *distributes* the assets to interested noncontributing team members such as programmers/developers.

The DFD's upstream asset is the TA. Its downstream assets are the Data Model and Code Module Inventory.

Data Model / Entity Relationship Diagram

> A structure organizing company and
> customer information into tables,
> columns, and rules for data integrity.

The Data Model (also called the Entity Relationship Diagram, or ERD) describes the organization—the structure—of a database (a later asset manages data *values*). The Data Model is often the technology asset that generates the most widespread interest. Many employees outside a project team talk about data and want access to it. An earlier section presented the project-independent asset Voice of the Customer and Seller, which often contains unstructured data. This Data Model technology asset is structured data used daily in business operations. The Data Model is subject to change in every project.

By extension of the general interest in databases, a typical project team also gives the Data Model a lot of attention. Database administration is a well-established profession and job title. Innovation professionals help their team by knowing a few fundamentals about the database and its asset, the Data Model.

The basic ingredients of a Data Model are tables, columns, and formats. A simple Data Model might have one table that resembles a spreadsheet of rows and columns. If a business's operational data is complicated, its Data Model requires dozens or hundreds of tables. Every table contains columns—individual data fields. Some tables have a handful of data fields, and others can contain dozens of columns. The Data Model assigns a format to every data field to define its maximum length and whether it contains alphabetic characters or strictly numbers (which allows software to perform calculations).

A high volume of data translates to a table with a high number of rows. A more advanced aspect of a Data Model—called the "primary key"—focuses on rows instead of columns to prevent duplicate data. Duplicate rows are a problem for software because code can't tell which of the duplicate rows to use. Process flows with strong governance reflect an understanding of what qualifies as duplicate data and contain process steps that prevent duplicating data. Data Models also prevent duplicate data by defining a primary key. A table's primary key specifies the data fields that qualify as a unique row of data to prevent duplicate data.

Another advanced aspect of a Data Model—called a "foreign key"—prevents orphaned data by governing relationships between tables. Many data fields appear in multiple tables, and data *values* appear in multiple tables.

A foreign key defines the fields and tables to allow and prohibit certain data values. The foreign key designates parent tables that allow and prohibit data values in children tables.

Data for a sports team can illustrate how a Data Model uses primary and foreign keys to prevent duplicate and orphaned data. If a sports league has two players with the same first and last names, a primary key (for a table "Players") with only those two fields sees rows for these players as duplicates. To avoid this, the Data Model adds a third data field—middle names—to the primary key so code can tell the two players apart. The same table might include another primary key containing fields "Team" and "Jersey number" to prevent any team from having two players with the same jersey number.

In this sports league's Data Model, to prevent orphaned data, tables called "Team" and "Position" are parents of a table called "Player." One foreign key requires that before code assigns a player to a team "Berlin Tigers" in the child table "Player," the team "Berlin Tigers" must exist in the parent table "Team." Another foreign key requires that before a player is assigned the position "goalkeeper" in the child table "Player," the position "goalkeeper" must exist in the parent table "Position." Database tools govern not only valid sequences to add data values but also valid sequences to delete data values, such as deleting the Berlin Tigers or goalkeeper from parent tables.

Data professionals and software developers execute Five Verbs for the Data Model. A junior database administrator *drafts* the Data Model's additions and changes for a project. Experienced data architects and software developers working on related code changes *review* and *revise* the Data Model. A data expert with a broad understanding of the company's technology actors *approves* the Data Model. The junior database administrator *distributes* the assets to interested noncontributing team members such as software developers.

Upstream assets of the Data Model are the Use Case Definition and the Data Flow Diagram. Its downstream assets are the Report Inventory, Data Definition Language, and Data Manipulation Language.

User Interface

> The selection, arrangement, and format of a technology actor's features that govern what the customer sees, smells, tastes, hears, and touches.

A conventional innovation methodology treats a user interface (UI) as the structure that strictly serves a human's sense of sight on an electronic device. A people-centric methodology envisions the UI as a structure that might serve all five senses of sight, smell, taste, touch, and hearing.

A conventional UI has data fields on the screen of an electronic device such as a laptop computer, grocery store self-checkout, or airport kiosk. Interaction—the exchange of information—between human and technology actors is two-way. A sloppy, uncaring technology actor barely provides instructions, validation, and confirmation of progress toward whatever common goal the actors share. A disciplined, empathetic technology actor provides instructions, validation, and confirmation of progress toward the common goal.

When someone entering data makes a mistake, a good UI (like a webpage) gives helpful, unambiguous error messages. The best error messages help users enter valid information on a second attempt. A good UI governs the customer experience and *content* according to the customer's *context*. It also works hard to minimize data integrity problems, rework for customers, and surprises for any stakeholders.

An expanded vision of a UI encompasses the human senses of taste, smell, hearing, and touch. Restaurants and makers of children's medicines care about customer perceptions of taste. Automated voice interactions via phone, retail shops, and large sporting events care about customer perceptions of sound. Amusement parks, museums, and bedding manufacturers care about customers' tactile perceptions. Casinos and hospitals care about customer perceptions of smell. Technology can play a role in managing expectations and perceptions for all five human senses.

Technology-enforced validation in a UI has a different cost profile than human-enforced validation. The technology has a high upfront cost to build (within a project) and a low ongoing cost to operate (post-project).[38] Human-enforced validation has a low upfront cost (no project cost) and a high ongoing cost to operate (significant incremental wages post-project). A disciplined, empathetic project team mindfully chooses the combination of human and technology actors to govern the customer experience cost-effectively.

Upstream assets define the degree and nature of intimacy between humans and technology, and the UI abides by the level of intimacy. Some UIs communicate one-way. Electronic food menus rotate displays of sandwiches, desserts, and weekly specials whether or not humans look at them. In contrast, many sensors listen, watch, and record humans but share nothing in return. Intimate, context-sensitive UIs might detect a customer's level of expertise and adjust according to whether human actors are novices or experts. A UI might detect whether to "trust" a human, giving them high autonomy and positive incentives. Or a UI might decide to "distrust" a human, restricting their options and explaining the negative consequences of bad-faith behavior.

High intimacy within a UI context can create a multitude of privacy concerns because some UIs are covert in their monitoring. Examples of overt monitoring include a website displaying, "Click here to accept cookies," and video meeting software alerting users that "This meeting is being recorded." Examples of covert monitoring include surveillance cameras, eavesdropping devices, and smoke detectors.

Professions such as graphic design, interior design, and user experience are well suited to contribute to Five Verbs for a UI. These specialists *draft* the UI. Sales or customer-facing employees *review* and *revise* the asset. The project sponsor *approves* the asset. One UI specialist *distributes* the asset to interested noncontributing stakeholders.

---

38 Parameterization–detailed in the Use Case Definition asset–is a great example of technology-enforced validation.

The upstream assets of the UI are Storyboards. Its downstream assets are the Code Module Inventory, Data Manipulation Language, and Report Detail. Expect to build the UI and the Data Model concurrently because the most important decisions within each govern less important decisions in the other.

Detailed Design

The second group of technology assets comprises detailed design (DD). These assets earn the DD title simply because they are neither process nor build assets, and their upstream assets are also design in nature.

The primary goal of upstream assets, such as process flows and the Use Case Definition, is *context management*. The primary goal of DD assets is *content management*. When someone considers certain innovation work so detailed as to be "in the weeds," they might be referring to DD assets.

A typical project team is highly familiar, competent, and disciplined with DD assets. Employees assigned DD work typically have the strong attention to detail the assets need. DD assets are the lowest-risk assets, but deliberate neglect prevents work on downstream technology (i.e., build and test assets), so cutting corners is impossible. Even innocent errors and inconsistencies are easy to detect because downstream assets must be congruent with every upstream asset. And fixing small errors that originate in DD assets requires minutes or hours.

The most sophisticated judgment among DD assets is aiming for a middle ground between the extremes of dense and trivial. Mindful design avoids putting too many or too few data fields in one place, such as a webpage, database, or report. The approximate middle is best because the extremes hurt the speed, quality, and ease of the customer experience. Precision in the middle ground is forgiving. Compared to the extremes, the middle ground is "elastic" and can accommodate changes in future projects.

Because DD assets inevitably have zero ambiguity, they are the simplest assets in innovation. More so than upstream assets, they make excellent assignments for junior employees.

Code Module Inventory

The titles and description of each
manageably sized section of code.

Before building the Code assets, a well-organized innovation team builds a Code Module Inventory (CMI). A CMI resembles the table of contents of a book or a theater program that lists every scene of the show.

A starting point for a CMI is the suite of Use Case Definitions. UCDs are well organized and modular, and they map to process flows. In the simplest scenario, the CMI maps one-to-one to the number of use cases in the UCD asset. But if the length of some code warrants more than one Code module, it's normal for a CMI to contain more modules than the number of use cases in the UCD.

Occasions to justify more Code modules than use cases include the *paths* defined in the UCD. UCDs capture the use case's primary, peripheral, parallel, contingency, and error paths. Any of these paths might justify their own Code module.

UCDs reflect the sequence and hierarchy of neighboring use cases. A diligent CMI does the same. The CMI asset is not a place for creativity or for diverging from the structure approved in upstream assets. Therefore, it might be sufficient to append the UCD with the names of Code modules instead of creating a separate, detached asset.

One goal of the CMI is to help divide code assignments among multiple developers to keep the duration of the build phase low. Another goal is to reduce the risk of large modules becoming too cumbersome to revise in the future. It's wise to avoid overburdened Code modules; modules with countless lines of code are prone to include inefficient and extraneous content, which gives rise to "spaghetti code," and large modules are more susceptible to multiple developers wanting to revise them simultaneously. A third goal of the CMI is to minimize redundancies, gaps, and waste in code. It's safer and easier to delete redundant and unused code by deleting an entire modestly sized module than by deleting a fraction of code in a large module. As the project team partitions code into modules, it should remember that future developers will find reasons to add, split, and merge modules.

Many programming languages have layers that foster modularity. Software is typically layered according to its proximity to the underlying operating system. Code near the operating system is rarely revised, and so it makes sense to partition it so that a typical developer does not access it. Software can also be layered according to its proximity to the user interface (UI). Code closest to the UI is called the presentation layer. Code closest to files or databases is called the data access layer. Code in the middle is the business logic layer. Whatever approach your company chooses, keep in mind the goals of modularity: low waste, high speed, and ease of isolating and revising code.

Because the CMI contains a small amount of new information and does not require developer skills or experience, a general administrator can *draft* the CMI. Junior and senior developers assigned code for the current project *review* and *revise* the asset. A senior software employee *approves* it, and the administrator *distributes* it to developers likely to revise the code.

The upstream assets for the CMI are the UCD, Technology Architecture, Data Flow Diagram, and User Interface. Downstream assets are Data Manipulation Language and Code.

Data Definition Language

Code that creates or
alters data structure.

The Data Model gives the project team a visual representation of the database, structure, and relationships. However, a database management system (DBMS) cannot interpret a revised Data Model itself. A DBMS needs literal instructions to create and change the database. These literal instructions are called Data Definition Language (DDL). DDL and two other detailed design (DD) assets (Data Manipulation Language and Report Detail) comprise the more common term SQL (Structured Query Language) that refers to all of them.

The instructions to create or alter a database are limited. They are:

- Add or remove a table.

- Add or remove a data field; for example, a column of the table such as phone number, email address, or backup contact.

- Change the length of a field (e.g., from 10 characters to 20 characters).

- Change the primary key of a table (defining what qualifies as duplicate data).

- Change the foreign key of a table (defining what qualifies as orphaned data).

DDL has a uniquely limited time relevance. A team uses a snippet of DDL once per environment for a single project, never to use it again. A team executes DDL instructions for their immature Development or Sandbox environment, then their Test environment, and finally their Production environment—and then never again. It makes no sense to repeat DDL execution within a single project or in a future project.

A project team must be cautious about whether DDL impacts existing data or triggers changes to other technology assets. Reducing field length can truncate some data values. Making a primary key stricter by changing the primary key from three to two fields ignores whether existing data would newly qualify as duplicate data. Making a foreign key stricter ignores whether existing data would newly qualify as orphaned data. Problematic DDL can cause a user interface (UI) to see data as ambiguous and cause code to crash.

A junior database administrator *drafts* DDL. One experienced data administrator can *review* and *revise* it. A data expert with a broad understanding of the company's technology actors *approves* the DDL. What qualifies as *distributing* the asset is executing the DDL against the DBMS to put the changes into effect in the respective Dev, Test, or Production environment. The upstream asset of DDL is the Data Model. The downstream asset is a Data Conversion Script.

Data Manipulation Language

> Code that adds, updates, or deletes data values.

Just as a database management system (DBMS) needs instructions for the database *structure*, it needs instructions to add, update, or delete data *values*. These instructions are called Data Manipulation Language (DML).

The first of three DML instructions is the "insert" statement. This statement adds one new row to a single database table. An insert statement specifies the relevant table, lists every field, and lists the variables (populated by code) that store the data values. Here's an example with hard-coded values:

Insert into Race-Registration (Event-ID, Name, Age)
Values ("00129", "Pat Nwajei", "33")

The second DML instruction is the "update" statement. This statement changes the values of one or more existing rows. Like the insert statement, the update statement specifies the relevant table, lists the fields to update, and lists the variables that store the data values. With rare exceptions, the update statement also specifies conditions to update a fraction of the table. For example:

Update Race-Registration
Set Age = "43"
Where Event-ID = "00129" and Name = "Pat Nwajei"

Although an insert statement always impacts a single row, an update statement can impact more than one row. If the example above omitted the "where" clause altogether, the statement would set the value of Age to 43 for every row in the table.

The third DML instruction is the "delete" statement. This statement deletes one or more existing rows. The delete statement specifies the relevant table and the conditions of the rows to delete. For example:

Delete From Race-Registration
Where Name = "Pat Nwajei"

This statement removes all rows for this name regardless of Event-ID.

One version of each of these three kinds of DML statements is usually best for each database table. Disciplined data professionals avoid creating multiple versions. Over several projects, teams revise the DML to add and remove data fields, so teams have less work changing a single version of DML instead of multiple versions scattered around countless code snippets.

DML statements depend entirely on the Data Model and are usually not specific to a single Code module. This enables a junior database administrator to *draft* DML. One experienced data administrator can *review* and *revise* it.

Assign a single developer who will use the DML in their code to *approve* the DML. Similar to Data Definition Language (DDL), what qualifies as *distributing* the DML is placing it where code can access and execute it.

The upstream assets of DML are the User Interface, Data Model, and Code Module Inventory. The downstream asset is a Data Conversion Script. It's natural to document DML after DDL, but a team might choose to document them concurrently.

Report Inventory

A list of reports, assignments, timing, and value.

Reports pervade business. What accompanies this rampant thirst for information is the low personal penalty for waste in reporting. That waste increases a company's cost and risk of information oversharing. A diligent Report Inventory helps to reduce waste and improves the governance of report generation. The asset reduces noise, cost, and risk.

A typical company does not mindfully manage a Report Inventory. Reporting requests are decentralized and countless, and they create information sprawl. This sprawl risks exposing sensitive information to individuals who lack legitimate business reasons to have the information and who might misuse it. A Report Inventory imposes discipline and mitigates the risk of misuse and a data breach.

For completeness and clarity of report specifications, every project should standardize certain information for every report. Every report has a title, description, audience or destination description, and requestor. Disciplined reports specify the date the report expires (unless renewed), their storage approach, and their specific purge date. Assign a senior point person to make decisions about fulfilling the report and a junior point person for tasks related to fulfilling the report.

Every report captures its dependency—a predecessor step in process flows. Describe the report's format or "visualization" such as all text, graphs, or interactivity (click-down) capability. Describe its expected length or volume of information.

Rate every report as high, medium, or low in its similarity to other reports. Highly similar reports are candidates to merge. Rate every report for its laboriousness (estimating the effort or duration to create). Laborious reports are candidates to automate or expire. And, finally, evaluate the report's net value, that is, the benefit of the information versus the effort to produce the report. Low-net value reports are candidates for expiration.

Below is a template for a Report Inventory.

Report Information	Report 1	Report 2	Report 3
Title			
Description			
Audience / Destination			
Requester			
Frequency			
Expiration or Renewal Date			
Storage Approach or Purge Date			
Senior Point Person Fulfiller			
Junior Point Person Fulfiller			
Dependency / Predecessor			
Format / Visualization			
Typical Size / Length			
Similarity to Existing Reports (High, Medium, Low)			
Laboriousness/Effort to Produce (High, Medium, Low or Number of Hours)			
Net Value vs. Effort to Produce "Earn vs. Burn"			

The asset has two levels in the execution of Five Verbs. For each report, the report requestor *drafts* a report-specific column. The junior point person *reviews* and *revises* it. The senior point person and the requestor's manager *approve* it. The report requestor *distributes* the single report information to the report's audience. Contributors to single reports need minimal information from the rest of the Report Inventory asset—likely limited to reports that rate similar to their own report.

For the entire asset containing dozens or hundreds of reports, assign a junior employee with a sensitivity clearance to *draft* every version (collating the columns). A database administrator (junior employee) and a database architect (senior employee) *review* and *revise* the asset, focusing on reports that involve them. Senior Operations, Compliance, and Information Security employees *approve* the asset. The junior employee with clearance *distributes* the asset only to the same approvers.

Upstream assets include the Scorecard, Future State Process Flows, and Data Model. The downstream asset is the Report Detail.

Report Detail

> Code that retrieves data from a database for logical processing, calculation, or display.

The final detailed design (DD) asset involves retrieving and presenting information from a database to a stakeholder. This asset deserves the most attention when the audience and information volume are large. These scenarios are captured in the Report Inventory, and the retrieval specifics are captured in this Report Detail asset.

Whether the volume of information to retrieve is large or small, the language is still the same, that is, SQL. The language for information retrieval is a "select" statement. Some code aims to retrieve a single row of information. Retrieval of a single row is likely intended for display via a user interface (UI). Here is an example of retrieving a single row of information using the example of an Outdoor Adventures registration:

> Select Age, Reg-Date, Amount-Paid
> From Race-Registration
> Where Event-ID = "00129" and Name = "Pat Nwajei"

Some code aims to retrieve a high volume of information. Here is an example of retrieving multiple rows of information. This query is a report containing the names, ages, and amounts paid of everyone registered for a single event.

> Select Name, Age, Amount-Paid
> From Race-Registration
> Where Event-ID = "00129"

In contrast to the low variability for Data Manipulation Language (DML; i.e., single statements for insert, update, and delete instructions), the variability of select statements can be very high. Every UI report is interested in customized information.

Report Detail is a unique asset because its output (the content and multiple rows of the report, executed within a process flow post-project) often impacts more than one project-independent asset. Report content might populate a lagging metric in the Scorecard, or it might generate ideas for the Change Log or expose a question, issue, or risk that belongs in the Parking Lot. When writing the Report Detail, include the Report Title so it cross-references the Report Inventory.

SQL for code retrieving a single data row is a low-risk asset. A junior developer or database administrator *drafts* the SQL. They consult one other similar junior employee to *review* and *revise* it. An experienced data architect or software developer *approves* it. Similar to the Data Definition Language (DDL) and DML assets, what qualifies as *distributing* this asset is a junior database administrator placing the SQL where code can access and execute it.

SQL for code to retrieve multiple rows of data is a report. Information in a report can be sensitive; changing the SQL can be contentious. Executing Five Verbs for Report Detail copies how it's done for the report's column in the Report Inventory, that is, the report requestor *drafts* the SQL, a database administrator *reviews* and *revises* the SQL, and a data architect and the requestor's manager *approve* the SQL. The junior database administrator *distributes* the asset by placing the SQL where the report runs according to the Report Inventory (i.e., adhering to triggers and applicable manual labor).

The upstream assets for Report Detail are the User Interface, Data Model, and Report Inventory. Likely concurrent assets are DDL, DML, Code Module Inventory, and Code. The downstream project-specific asset is Data Conversion Script. The downstream project-independent assets are the Scorecard, Change Log, and Parking Lot.

Build

Design assets give unambiguous guidance for Code and its peripheral assets in the build phase. On par with the upstream group of detailed design (DD) assets, build phase assets are another example of getting into the weeds of innovation work. Code is the centerpiece technology asset that does the heavy lifting for humans, and the other build assets move code and data from immaturity to customer readiness.

A typical project team doesn't neglect build assets. Legacy methodologies ensure this and even elevate Code in importance at the expense of more foundational assets in shaping the customer experience. Build phase assets are mature due to decades of software development.

Build assets are exciting because developers are finally creating and improving the technology actors that perform the expectations specified in all the preceding assets. Code is what technology actors execute countless times to serve the customer and improve their lives. The other build phase assets make the build process easier, quicker, and less vulnerable to mistakes.

For better and for worse, build assets contain minimal creativity. Decisions about the organization and goals of each build asset have already been made. An exception can arise if a developer shares a concern about a design asset and recommends revisions.

It's impossible to work in innovation and dismiss technology or code, so build assets rarely have skeptics. However, code elicits skeptics on how to apply the Five Verbs framework. Developers don't use the term *draft* for their code. They use the word *code* as a verb and a noun. Therefore, developers are prone to reject the rigor, standardization, goals, and ideas behind the Five Verbs framework.

But Five Verbs applies to Code because developers *draft* Code, *review* each other's Code, and *revise* Code. Someone *approves* Code (explicitly or implicitly) before someone promotes (*distributes*) it to a more advanced environment, such as a Test or Production environment. Like every other asset in teamwork, Code without Five Verbs is vulnerable to mistakes, siloed thinking, poor alignment, and poor accountability.

Because of the software-centric methodologies, the build assets already have high durability. It's common for forward-thinking developers to include comments in their work—explanations and instructions to those who will change the assets in the future. Comments aim to reduce work for future developers and make their job easier.

Code

> The script for technology to perform logic, manipulate and store data, exchange information with other actors, and provide products or services to humans.

Code contains instructions for the work humans delegate to technology. Technology actors don't intuitively know what to do. Humans (developers) must give precise instructions for everything they want technology actors to do. Like human actors in theater, technology actors need a script, and that script is Code.

Companies and innovation teams rarely neglect Code—it's more common to hear about and experience a shortage of developers instead. Code is universally recognized as a path to automation, innovation, customer value, and profit. If anything, companies are overly eager to code and build it *without* the prerequisite assets. But when companies build it with customers in mind, Code takes on the eight Ds type of work: dull, dirty, dangerous, difficult, demanding, demeaning, delicate, and dear. Developing Code is a great career path.

Logistically, work on the Code asset starts after detailed design (DD). But a developer gets their specific guidance for drafting Code from five particular assets. Code's first responsibility is to obey the Future State Process Flows (FSPF). Process flows shape the dialogue among all actors, often a single human actor and a single user interface (UI). Process flows tell the Code every path (primary, peripheral, parallel, contingency, or error) and every step to build into its script. Process flows show where information sharing happens.

Code obeys the detail found in Use Case Definitions (UCDs). UCDs explain the intended number of process steps, handoffs, and how long each actor should wait for another actor. They also explain the data inputs and outputs of every code module. UCDs govern which numbers are hard-coded versus parameters presented to a human that can change.

Code obeys Storyboards. Although Code is "robotic" (and might even reside literally inside a robot), it aims to fulfill the emotional journey the Storyboards define. Code supports the pain and exhilaration of the humans in the dialogue.

Code obeys the UI. UI assets specify what information to collect from and share with humans. That information might be limited to keystrokes, text, and images on a webpage, or it might span any of the five senses of sight, smell, taste, touch, and sound.

Code also obeys the Data Model. The Data Model specifies what information the technology aims to store. Information to be stored significantly overlaps the information exchanged via the UI, but Code commonly references additional information not shared in the UI. Code uses Data Manipulation Language (DML) to insert, update, and delete data and uses Report Detail to retrieve data for calculations and to share with the user.

Code is a unique asset because, during the work, the word "can't" might appear. A developer might conclude that the code cannot accomplish an expectation set in an upstream asset. The code might not achieve the expected number of handoffs, nor detect the customer's emotion, nor force the human to exchange information. A developer should escalate this promptly, so the team considers options and potentially resets expectations before approving the code and proceeding with the project.

Whenever Code is simple enough, a junior developer *drafts* it. More experienced developers *review* and *revise* it. A senior technology employee *approves* it. What qualifies as *distributing* Code is placing it where the technology platform can access and execute it. Upstream assets for Code are the Data Model, User Interface, and Code Module Inventory. Downstream assets are Migration Script and Test Conditions.

Migration Script

> The inventory and instructions to move technology actors among environments.

Most companies partition their customer-facing technology from technology that is not yet ready for customers in separate environments. A project team moves technology actors among these environments. When a team expects to repeat this migration enough times, it makes sense to build a Migration Script instead of moving technology components manually.

Every company that has significant customer-facing technology has an environment it calls "Production" and a non-customer-facing environment that is commonly called "Development" or "Sandbox." In between these environments, large companies have one or more environments named for the training or testing they support.

In every project, a team builds new Code in the "Dev" environment. Whether the amount of new Code is large or small, the team tests it. Once team members are confident in that Code, they move it (and associated technology assets) to a "Test" environment. The team then tests the Code and how well it works with unchanged and mature Code. Once the team is confident in the small Code modules and their integration into the entire customer experience, it moves the Code and other technology assets to Production.

Moving technology assets repeatedly and piecemeal is dull and error prone. The Migration Script standardizes the inventory of technology actors to migrate among environments. The typical assets to migrate are User Interface (UI), Data Definition Language (DDL), Data Manipulation Language (DML), Report Detail, Code, and Data Conversion Script. A large project likely includes dozens or hundreds of technology assets in its Migration Script.

A typical developer understands problems their code might have integrating with other code and processes. Reducing the laboriousness of migrations allows for more attention to be paid to verifying a successful migration and testing how well the newly arrived code integrates into its new environment.

The primary developer who drafted the Code is the best candidate to *draft* the Migration Script. The developers who reviewed and revised the Code are good candidates to *review* and *revise* the Migration Script. A senior technology employee *approves* it. The primary drafter *distributes* the script to an administrative employee who executes it, moving the technology actors to their next environment. Upstream assets of the Migration Script are the DDL, DML, Report Detail, and Code. Downstream assets are the Actual Results and Deployment Plan.

Data Conversion Script

Scripts to add or update
a large volume of data.

Some projects have a solitary goal of populating or changing a large amount of data. Lone goal or not, when the volume of data is high enough that adding or changing data values manually via the user interface (UI) is unacceptably laborious, unrealistic, or plain impossible, the project team builds a Data Conversion Script.

A typical company has data professionals familiar with "extract, transform, and load" (ETL). A Data Conversion Script is a more generic form of ETL. Both aim to change a high volume of data and avoid time-consuming, error-prone manual data entry.

One form of data conversion *adds* multiple rows to a database table. A company experiencing a merger might merge similar data from two companies into a single database and corresponding tables. A company might have received contact information for hundreds of prospects in a file format and wants to store this information without manually populating the database. A script that adds hundreds of rows reads every row, formats it to be compatible with the database, and executes an insert (DML) statement against it.

A second form of data conversion *updates* multiple rows of a database table. When a project adds or changes a field in a database, it might want to populate the field in a single event instead of piecemeal. Updating the field might be a single update (DML) statement with a single data value.

A third form of data conversion preserves some data values and updates other parts. In this scenario, the script reads each row, makes the change, and outputs the changed data field in the intended format, with the Data Conversion Script updating one row at a time. Scenarios exist where a script parses a single data field and splits the information into two separate fields (e.g., "Name" into "First Name" and "Last Name"). Some of these intricate data conversions involve a file instead of a database, for example, converting a file from an old format to a new format.

Although Data Conversion Scripts are valuable for avoiding mistakes in manual data maintenance, the script itself could contain a mistake or a detail with unintended consequences. Mass data changes are risky because reversing the changes could be painstaking work if the results aren't what the team intended. Because of this risk, creating the script itself warrants caution. The project team should build and test the script in the Dev and Test environments and examine the results before executing the script against Production data.

On occasion, the data conversion is significant enough to be the centerpiece of an entire project. Such a project could relate less to the customer or employee experience and more to the data values so operations can execute in some new and desired way. After executing the script, team members should inspect the data to verify that the values are what they expected.

A detail-oriented business analyst, an Operations employee, and possibly a salesperson are appropriate team members to *draft* the Data Conversion Script. At a minimum, an experienced data architect *reviews* and *revises* the script. Several project stakeholders may want to *review* the script. The project sponsor *approves* the script. What qualifies as *distributing* the asset is executing the script in each environment. The upstream asset for a Data Conversion Script is the Data Definition Language. Its downstream assets are the Actual Results and Deployment Plan.

Deployment Plan

> The script for the project team and Operations employees to place hardware, software, and data into the Production environment for customers to use.

When a project's technology assets approach completion, the project team organizes a schedule to move the actors into place, enabling the new customer experience. A Deployment Plan is the schedule for placing hardware, software, and data into the Production environment.

A typical project team has a Deployment Plan. Companies allow the plan to differ in scope and emphasis, but the coinciding event (a go-live event) and asset (Go-Live Announcement) are typically high-profile, so practically all project teams formalize deployment assets appropriately. The sequence of steps is usually scripted, and the pace of work is conservative.

An approximate sequence of steps takes existing hardware and software offline, then deploys new hardware, new software (executing a Migration Script), and data (executing a Data Conversion Script). The plan brings new hardware and software online and performs a combination of new and legacy processes (casually called "smoke test"). The plan instructs that the team monitor performance until it is confident in the stability of new processes and technology. Every step of work outlined in the Deployment Plan contains the name of the assigned employee executing the work and the estimated start and completion times of their work.

If all deployments went smoothly, that would be enough, but because not every deployment is smooth, every Deployment Plan needs a rollback procedure. These steps are necessary to restore the old hardware, software, and data. Another tactic for careful monitoring is called "hypercare." Hypercare dedicates team members to monitor performance full time and field inquiries for the first few days or weeks of the new processes and technology.

The Deployment Plan should consider a few other concepts to keep risk low. Standardization across projects keeps risk low by reducing the chance of an oversight. The team can mitigate risk for large, complex projects by executing the Deployment Plan for other environments before executing it for its Production environment. The team might stagger the availability of the new processes to customers over multiple go-live events. The company reduces risk by having a handful of customers use the new processes and technology until confidence is high that the new capabilities can successfully support more users. The company might also have the luxury of temporarily hosting the old and new experiences simultaneously, then can make the old experience unavailable once the new experience is stable.

A Deployment Plan requires expertise of developers, data administrators, and operations. A developer, database administrator, and Operations employee—all junior—can *draft* the Deployment Plan. Experienced developers, technology architects, data architects, and Operations employees *review* and *revise* the asset. Senior technology, data, and Business Operations employees and the project sponsor *approve* the asset. The junior Operations employee *distributes* the Deployment Plan to team members with assignments in the asset, and to smoke testers and the hypercare team.

Upstream assets for the Deployment Plan are Migration Scripts, Data Conversion Scripts, Actual Results, and training materials. Execution of the Deployment Plan coincides with the distribution of the Go-Live Announcement. The Deployment Plan's project-specific downstream asset is the Closure Report. Its project-independent downstream assets are Scorecard, I Like I Wish I Hope I Wonder, Change Log, and Parking Lot.

Test

An innovation team should not expect its technology and related processes to work correctly the first time. A disciplined team performs structured tests on processes and technology before releasing new customer experiences to stakeholders. Diligent testing maximizes the quality of the customer experience and minimizes risks of customer frustration or backlash against the company.

A typical team performs some unstructured tests on new Code assets as well as some informal user acceptance testing. But many teams shrug at formal test plans, exhaustive tests, and the use of fresh employees to maximize the rigor of tests. This neglect squanders the value of testing and increases the risk of bad customer experiences. A disciplined test phase builds assets that govern thorough testing and systematically involve multiple testers and fresh perspectives. Disciplined testing finds problems before customers find them for you.

Testing is not a single exercise. Rigorous testing has three, four, or even five phases: Benefits Realization, Operational Readiness, User Acceptance, System Integration, and Unit Testing. Test phases *execute* in a specific sequence, and each phase's test *planning* work can—and should—happen in the reverse order. Each test phase holds certain upstream assets accountable for what they said they would do for the new customer experience. This table lists five recommended test phases in the sequence of test planning work and maps each test phase to its corresponding upstream assets.

Test Phase	Upstream Assets
Benefits Realization	Project Charter Elevator Pitch
Operational Readiness (often bundled with Performance, Penetration, or Stress)	Technology Architecture
User Acceptance	Customer Lifecycle Model Storyboards Use Case Definition
System (or Technology) Integration	Code Module Inventory Data Flow Diagram
Unit	Process flows User Interface Data Model Data Manipulation Language Report Detail

One test asset, the Test Approach, applies to all test phases and governs all other test planning and test execution assets. All other test assets require versions for every test phase. Every test phase needs its own test *planning* assets and builds them in this sequence: Test Conditions, Test Cycles, Test Scripts, and Test Data. Every test phase requires its own test *execution* assets and builds them in this sequence: Actual Results and Defect Log.

The maturity of testing has shaped the profession of "Quality Assurance" (QA) to apply to tangible products and software. QA is a common function in innovation organizations. Every project has project-specific team members, even though some rarely work on projects. QA professionals provide a valuable cross-project perspective and can coach every aspect of testing. The "QA team" is often synonymous with the "testing team."

Test assets shape a culture of discipline through the project team's *vigilance* in verifying the *quality* of processes, technology, and customer experience before sharing them with customers. These assets shape a culture of empathy by asking employees to be *vulnerable* about—and not avoid or suppress—problems, mistakes, and small *rejections*. Test assets set the expectation that finding problems is the primary reason for the work. Fixing problems is a form of *rehearsal*.

Test Approach

> The high-level expectations for test planning and test execution for a new customer experience.

Just like a Training Approach sets high-level expectations for training work and might govern multiple projects, a Test Approach sets expectations for testing work and can apply to multiple projects. A Test Approach poses the same questions to every project, encouraging consistent answers but allowing projects to answer differently.

A typical company forms a Test Approach, but only casually in the minds of a handful of employees passionate about testing. This ambiguity hurts the speed and quality of testing. Diligent testing professionals—often QA—are enthusiastic about setting expectations, minimizing surprises, and formalizing a Q&A that serves testing needs for every project.

The Test Approach sets three specific expectations at the level of the *project*. Specify the number of test phases, the names of the test phases, and the number of test environments (two test phases might use a single environment). The Test Approach sets two expectations at the level of each *test phase*. Describe the test *environment* needs (e.g., hardware, operating system software, network configuration, database setup, licenses, automation tools). Summarize Test Data expectations (e.g., dozens of events, hundreds of customers, thousands of transactions).

Finally, the Test Approach default asserts that every test phase creates every other test asset. The Test Approach allows a project to opt out if it feels unnecessary and like overkill to create some asset. For example, a project conducting four test phases and only opting out of Test Cycles for the System Integration Test (perhaps considered a small test) includes this table in its Test Approach. Otherwise, this Test Approach accepts defaults to create all other test assets.

Phase Asset	Operational Readiness	User Acceptance	System Integration	Unit
Test Conditions	Y	Y	Y	Y
Test Cycles	Y	Y	~~Y~~ No	Y
Test Scripts	Y	Y	Y	Y
Test Data	Y	Y	Y	Y
Actual Results	Y	Y	Y	Y
Defect Log	Y	Y	Y	Y

A junior QA analyst can *draft* the Test Approach. A group of business analysts and QA employees—especially those likely to have assignments for other test assets—*review* and *revise* it. The project sponsor and QA leader *approve* it. The junior QA analyst *distributes* the Test Approach to all employees who contribute to test assets, including QA employees and testers. QA employees and the QA leader informally monitor Test Approaches across projects to minimize siloed decisions, inconsistencies, neglect, and overkill in test assets.

The upstream asset of the Test Approach is the Project Charter. The downstream assets are the Test Conditions for the test phase whose upstream assets are finished. For example, start Test Conditions for the Benefits Realization Test when the Project Charter is done. When the Code Module Inventory and Data Flow Diagram are finished, start Test Conditions for the System Integration Test.

Test Conditions

> Laundry lists of details to test
> as dictated by upstream assets.

Whereas the Test Approach spans test phases and even projects, Test Conditions are unique to a project and a test phase. In this step of test planning, inspect the relevant upstream assets and list what you believe needs verifying that it works as expected.

A typical QA organization builds Test Conditions but isn't careful about phrasing and organizing them. Haphazard Test Conditions are vulnerable to overuse of familiar terminology, personal bias, and lack of attention to detail. The pragmatic, disciplined approach to drafting Test Conditions is to focus on the relevant upstream assets and to scrutinize which errors a human or technology actor could make in performing the asset's specifications, i.e., in meeting the asset's expectations.

The nature of appropriate Test Conditions is the same across projects but different for every test phase because of different upstream assets. The table on the next page illustrates Test Conditions for every test phase.

To build Test Conditions, apply Five Verbs for every test phase. QA professionals have prominent assignments and serve as coaches to project-specific employees. Their contributions distribute workload, decentralize accountability and infuse testing skills among employees. The table on page 218 proposes assignments of Five Verbs for Test Conditions across test phases.

Upstream assets for Test Conditions are listed above in the Test introduction section for their respective test phase. Downstream assets are Test Cycles, also by test phase.

Test Cycles

> Bundles of Test Conditions to ease
> the execution of a large test exercise.

Test Phase	Sample Summary Test Conditions
Benefits Realization	Price of customer experience (revenue) Quantity of customer experiences (revenue) Upfront cost to provide customer experience (project cost) Marginal cost to provide customer experience (operating costs) Moments that matter Metrics that matter
Operational Readiness	Technology response time Simulate high volume of users/traffic Breach
User Acceptance	Actor list Number of steps Number of handoffs Duration to complete use case Effort to complete the use case Customer Lifecycle progress Customer sentiment
System Integration	Database contents (inputs and outputs of Use Cases) File contents (inputs and outputs of Use Cases)
Unit	Capture every data field (including parameters) Retrieve every data field Data field value limits Data values to trigger every logic branch Process governance

Sprawl can exist in many forms, including in Test Conditions. When a test phase has a high number of Test Conditions, a team can bundle them into portions of tests called Test Cycles. Test Cycles reduce risk and *rework*. They facilitate purposeful *retesting*.

A typical testing (QA) team keeps Test Cycles in mind for their largest projects and sees value in partitioning tests. Partitions encourage broad participation and reduce the intimidation of testing. Test Cycles provide clear starting points for testers and reduce a big testing exercise's real and perceived burdens.

Test Phase	Draft	Review and Revise	Approve	Distribute
Unit	Junior developer	Other junior developers, QA team	Experienced developers	Same as drafter Distribute to testers
System Integration	Junior technical architect	Experienced technical architect, QA team	Senior Operations employee	
User Acceptance	Junior business analyst	Junior customer-facing employees, QA team	Senior customer-facing employees and project sponsor	
Operational Readiness	Junior Operations employee	Senior Operations employees, QA team	Chief operating officer, chief architect, project sponsor	
Benefits Realization	Junior customer-facing employee	Senior customer-facing employees	Project sponsor	

A Test Cycle is a bundle of Test Conditions. A handful of Test Conditions in a test phase can't justify more than one Test Cycle. But if the test phase has enough Test Conditions to justify partitioning them, name every Test Cycle you have in mind and assign every Test Condition to one of the Test Cycles.

A tester *creates* a starting point for a Test Cycle when they pause executing a Test Script and immediately back up the database. A tester *uses* the starting point of a Test Cycle when they restore the data later and resume testing. Testers might feel constrained when they have only one or two places to start a test, but the workload and risks are low when they have a dozen options to execute a Test Script. After a certain defect is fixed, Test Cycles helps testers verify fixes since they likely can start retesting close to where they experienced the bug.

A team organizing Test Cycles must have a technology tool to configure and perform the backup and restore capability. A tester skilled in restoring the database gains confidence in testing, which improves their confidence in the process and in the technology they're testing. Armed with a Test Script, a tester can feel inspired to test and retest more rigorously. This raises the tester's accountability and ownership, as well as their trust in the test and ambition in what they test. Because of this, although Test Cycles appear administrative, they shape the culture for testing.

Another purpose of a Test Cycle is to be an interim asset between its upstream asset (Test Conditions) and one of its downstream assets (Test Scripts). A Test Cycle serves as a container for both. A Test Cycle contains unsequenced Test Conditions, and a Test Script contains sequenced Test Conditions. A Test Cycle might include more than one Test Script. To get a sense of quantity, a test phase with one hundred Test Conditions might have three Test Cycles (entry and exit points) and ten scripts. A test phase with one thousand Test Conditions might have thirty Test Cycles and one hundred scripts. The high quantity of Test Conditions and scripts in large projects complicates assigning tests, pacing tests, and measuring the progress of test execution. Test Cycles lower the number of moving parts to organize, easing this project management work.

So that diverse perspectives and skills shape test assets, for *draft, review,* and *revise* consider a straight swap of assignments with the people assigned for Test Conditions (as shown in the following table). QA professionals are likely to have a cross-project view of organizing Test Cycles, and project-specific professionals should contribute their project-specific perspective.

Test Phase	Draft	Review and Revise	Approve	Distribute
Unit	QA analyst	Junior developers	Experienced developers	Same as drafter
System Integration	QA analyst	Junior technical architects	Experienced technical architect	
User Acceptance	QA analyst	Junior business analysts Junior customer-facing employees	Senior customer-facing employee and project sponsor	Distribute to testers and add to the test automation tool
Operational Readiness	QA analyst	Junior Operations employees	Senior Operations employees	
Benefits Realization	Junior customer-facing employee	Senior customer-facing employees	Project sponsor	

Upstream assets for Test Cycles are Test Conditions for each test phase. Downstream assets are Test Scripts, also by test phase.

Test Scripts

> The sequence of steps for a tester.

To maximize speed, quality, and ease when executing tests and to reduce duration, effort, and waste, every disciplined project team builds Test Scripts. Test Scripts set up testers to verify the integrity of the upstream assets for each respective test phase.

But many typical project teams build a minimum of Test Scripts and cover a single scenario or path. Plenty of teams don't build Test Scripts at all and strictly try to break processes or systems in an ad hoc manner.

Test Scripts need attention to detail to maximize the chance that testing verifies customers' functions, governance, and freedom.

A diligent team builds enough scripts so that structured testing can confirm that processes and technology work as expected. Diligent Test Scripts show testers how to verify that all customer experience paths (primary, peripheral, parallel, and exception—as defined in process flows) function properly. Diligent Test Scripts include steps for testers to distinguish mandatory, optional, improvised, and prohibited actions and verify that they work as expected.

A diligent team cannot and should not build scripts for an infinite number of improvisations by a tester (or ultimately by the customer after go-live). Improvisations are literally "off-script."[39] Some testers like to go off-script, and unstructured testing on top of structured testing is valuable because it can expose positive and negative surprises.

Ideally, upstream assets address nefarious actors such as hackers. Whether or not they do, Test Scripts are another asset to test, mitigate risk, and proactively navigate a breach of processes or data. Testing a breach is called a Penetration Test[40]—often a subset of the Operational Readiness Test. Security is such an important topic that it necessitated the profession of "ethical hacking" and the job title Chief Information Security Officer.

Test Scripts promote cognitive diversity for the user experience. Executing a Test Script for the first time, a tester likely does not get creative or curious. But if a tester runs a Test Script enough times, they might get a sense of the limitations of the customer experience. If it is constructive to do so, a tester might experiment with the script and uncover a defect that none of the upstream assets anticipated. Test Scripts set the table for rigorous tests. Discovering and fixing these unanticipated problems before customers inevitably discover them avoids huge costs and crises. Experimentation can

---

39 Freedom for improvisation is most applicable in User Acceptance Test and mildly applicable in Unit Test. Other test phases have practically no improvisation.

40 Penetration Test is an authorized simulated attack to gain control of a company's processes and/or data; fixes reduce vulnerability.

even lead to "positive" surprises worth evangelizing because positive surprises raise customer interest and can ultimately increase revenues.

The first three columns of a Test Script resemble Current State and Future State Scripts. The drafter numbers each step and specifies the human or technology actor and their action. They reference the relevant Test Conditions (to minimize a script meandering into irrelevant tasks). They add Test Data that the script requires for the test to work. They specify an expected result when that result might be something other than merely the next script step.

The next column to populate in a Test Script is a reference to stoplight ratings (Red, Yellow, Green) in the Use Case Assessment. Stoplight ratings might pertain to the actor list, the number of process steps, or handoffs. They might pertain to the use case's duration or effort/laboriousness. In this column, reference the nature of the improvement, for example, in duration, effort, or handoffs.

In the Test Scripts of every project, specific steps improve a use case toward Green ratings.[41] For example, if five steps of a script are instrumental in reducing a use case's duration, write "duration" in the Stoplight Impact column for those five script steps. If twenty steps are instrumental in reducing the number of handoffs, write "handoffs" in the column for those twenty steps.

Finally, the Test Script template shown on the next page reserves three columns for the tester to populate when they execute the script: a column for the tester to note "pass" or "fail" (for a Test Condition or a single script step), a column for the step's actual result, and a column for the tester to reflect a defect number.

Many scripts' first step tells the tester to restore data to a particular state. Many scripts end by telling the tester to back up the database so it can be restored later and to resume testing from that point. Give every Test Script a name (it might match the use case it most closely matches) and cite the Test Cycle that contains its Test Conditions.

---

41 Stoplight impact applies to every project, but not every test phase. It's possible a project might only use this column for the User Acceptance Test phase.

#	Actor	Action	Relevant Test Condition	Test Data Needed	Expected Result	Stoplight Impact	Pass Or Fail	Actual Result	Defect #
1									
2									
3									

A peripheral benefit of Test Scripts is that they enable test automation. Many tools allow you to record Test Scripts so their playback can automate test execution. This accelerates a type of testing called "regression" testing, which aims to verify that new code or processes do not break unchanged code and processes.

In assigning Five Verbs for Test Scripts, consider the same assignments as for Test Conditions. Project-specific junior team members are best for *drafting* Test Scripts. Experienced employees and QA professionals are best for *reviewing* and *revising* them. Corresponding senior employees are best to *approve* them, and the junior team members *distribute* the Test Scripts to testers.

Upstream assets for Test Scripts are Test Cycles. Downstream assets are the Actual Results for each test phase. For each test phase, Test Data is the concurrent asset for Test Scripts.

Test Data

> Rows and columns of information in
> files and database tables that support
> each test phase, cycle, and script.

When a tester starts executing Test Scripts, files and databases are not empty but contain relevant information that allows testers to execute Test Scripts. This information is the asset Test Data.

Test Data can occupy many database tables and files. As a tester proceeds through a script, the Test Data changes, and verifying the success of a test requires inspection of those same database tables and files.

Some project teams casually grab data from another similar database with a high volume of data that resembles the data in their Production environment. This method has a few problems. This data might contain sensitive information that should not be seen or reside elsewhere. It is likely incompatible with what the new technology actors can use. Curating the data can be time-consuming, thus undermining the reason for a data grab. The data might not contain the characteristics needed to exercise new processes. It inevitably contains extraneous information that slows data inspection.

Curated Test Data has a handful of benefits. A thoughtful, uncluttered number of rows helps test planners and testers focus on what every test phase, cycle, and script needs to accomplish. The tidiness of curated Test Data can expose new Test Conditions, business logic, data combinations, and data relationships.

A project team starts identifying Test Data while it drafts Test Scripts. For the fictional company Outdoor Adventures, drafting a Test Script exposes that, to verify different Test Conditions, Test Data must include a certain number of events, camping sites, and participants. For example, a large project of a hundred Test Conditions could require ten events, thirty campsites, and thirty participants (three per event).

Not only does it make sense to draft Test Data while drafting Test Scripts, but it's also common to finish drafts of both assets simultaneously because the final Test Scripts can still identify Test Data needs. In the first drafts of these assets, scripts "lead" the data. Conversely, a plain inspection of Test Data can expose new Test Conditions and Test Scripts. For example, certain data combinations might influence a mandatory or prohibited action in a Test Script that a team realizes is important only while inspecting Test Data.

Test Data serves every potential starting point across test phases, cycles, and scripts. After team members load Test Data in an environment, they should back up the data to restore it every time a tester re-executes a Test Script. A Test Script changes data, and the team will likely want to back up that data to resume testing from that point among Test Scripts. Some back-ups should be layered, where a first backup involves one set of tables, and a second backup involves a partially overlapping set of tables for a different starting point for a script. For example, a project for Outdoor Adventures might establish a first data backup containing events, campsites, participants, and transactions. A second data backup might contain only *events* and *participants*, and a Test Cycle might restore these backups in succession to retest with data in a particular state.

The assignments of Five Verbs to Test Data resemble assignments for Test Conditions. This means project-specific team members *draft* curated Test Data. QA employees, who have a cross-environment, cross-project view of Test Data, *review* and *revise* Test Data. Senior employees and the project sponsor *approve* the Test Data. What qualifies as *distributing* Test Data is the QA team loading it where testers can easily access and restore it to start testing.

The upstream asset of Test Data is Test Cycles. The downstream asset is the Actual Results for each test phase. The concurrent asset for Test Data is the Test Scripts for each test phase.

Actual Results

> What the user sees on the User Interface or in the database after Test Script execution; what the tester experiences with all five senses.

Testers—employees who execute Test Scripts—also have assets to draft. Their detailed Actual Results append to the Test Script, and their summarized test results contribute to status reports.

Many testers limit their documentation to the next asset—the Defect Log. Bug descriptions and resolutions are critical information, but they are not evidence that the project team verified new processes, technology, and that customer experiences fulfill their missions. A diligent project sponsor mandates Actual Results to minimize post go-live legal and reputational risk and to maximize trust and confidence in the new customer experience about to launch.

Like the upstream test assets, Actual Results assets apply to every test phase. However, testers execute scripts and build Actual Results assets in the reverse order of test phases. A company with five test phases performs Unit Test, System (Technology) Integration Test, User Acceptance Test, Operational Readiness Test, and Benefits Realization Test. Every test phase should reach a GETMO pass rate before the next phase begins.[42] Every transition between test phases leverages the Migration Script and Data Conversion Script.

For drafting Actual Results, a tester's default approach is writing the word "Pass" for every script step that proceeds as expected, that is, where the actual result matches the expected result, and the tester has nothing more to report or reflect. They leave the columns Actual Result and Defect Number blank.

When an actual result differs from the expected result, the tester writes "Fail" and records what they see (or smell, feel, hear, or touch when the experience expands beyond a webpage). They enter this detail as a new item in the Defect Log, note the new defect ID (number), then record the defect ID in the final column of the Actual Results as a cross-reference. The tester has to judge whether it's possible and worthwhile to continue executing the script. The primary reasons to continue are to determine how severe the defect is and to discover additional defects, accelerating their fix.

A tester executes Test Scripts until no scripts pass, defects feel like they're "shadowing" an already-found defect, or defects prohibit the tester from proceeding. When the tester pauses executing a script, they share their Actual Results and the Defect Log with other team members so that fixing defects (reviewing and revising the relevant upstream assets) can proceed.

The conventional view of a defect says a problem exists with the process or technology (giving the benefit of the doubt to the test planning asset). But occasions exist when a problem resides in a test asset, such as minor errors in the Test Script or Test Data. A tester should consult others to determine which asset or assets warrant revising before resuming the test.

---

42 Typically, 95 percent qualifies as good enough to move on, but companies should identify their own standards for GETMO.

With every retest, a tester gains familiarity with the processes and technology and gains more confidence in the success of being on-script. A tester might then creatively explore how the processes and technology respond to behaving off-script. They could explore whether the process is childproof and safe from misuse, or whether the process contains any positive surprises. Both scenarios can be valuable to the project or to subsequent projects. The tester is uniquely close to the customer experience, so they're uniquely positioned to be creative, discover blind spots, and propose new ideas for innovation. Explorational testers record off-script Actual Results in the Test Script when those detailed results might justify follow-up. Testers might add explorational conclusions and enhancement ideas in the Defect Log or in project-independent assets such as I Like I Wish I Hope I Wonder, Change Log, and Parking Lot.

In addition to documenting detailed Actual Results appended to the Test Script, diligent testers also summarize Actual Results. For every testing session, testers report the following:

- Test Conditions attempted

- Test Conditions passed

- Pass rate (a simple division of these two numbers)

- Numbers of defects (newly discovered, fixed, and not yet fixed)

- Expected stoplight impact, that is, success or failure toward the project's value proposition

- Recommendations from any off-script exploration

- Whether testing feels sufficiently complete—GETMO to the next testing phase

Test phases with dozens of Test Conditions have a pass rate that progresses like an S curve (diagram below). Early in test execution, the pass rate increases slowly. At some point, the pass rate rises quickly, and near the end of a test phase, the pass rate plateaus as it approaches 100 percent. A pass rate of 95 percent is a common place for a team to declare GETMO for a test phase. A common motivational tool is to display this graph (Figure 11) for testers to monitor progress.

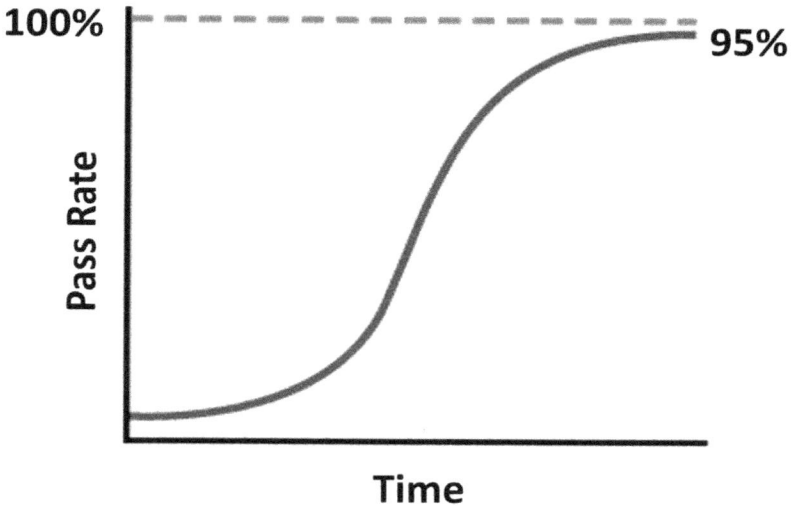

Figure 11. The pass rate of test conditions typically follows the path of an S-curve.

Actual Results are assets with little substance (two columns) but significant value. Every "pass" a tester captures is a micro-signoff—an approval—that a Test Condition works as expected. Actual Results represent the test's integrity and the tester's accountability. Actual Results also facilitate calculating the pass rate. For example, for a Test Cycle with five scripts containing fifty Test Conditions, eight passed Test Conditions is a pass rate of sixteen percent.

Healthy assignments for Five Verbs are similar but not identical to assignments for Test Conditions—involve the same skills and seniority, but not all the same people. A great candidate to serve as a tester—*drafting* Actual Results—is a junior employee not involved in any upstream test assets. Involve some fresh perspectives to *review* and *revise* Actual Results. The same senior employees (contributors to Test Conditions) and the project sponsor *approve* the Actual Results. A QA employee *distributes* Actual Results to stakeholders who want to boost their confidence in the new capabilities by seeing the test results.

Upstream assets for Actual Results are Test Scripts and Test Data. Its concurrent asset is the Defect Log. Downstream assets are the Actual Results for other test phases and a Go-Live Announcement.

Defect Log

> A repository of defects and enhancements that help a project go live with minimal problems in the processes, technology, and customer experience.

As testers execute Test Scripts, they capture Actual Results in two columns next to other Test Script detail. When testers find a problem, they capture information in a separate asset—a Defect Log—then cross-reference the defect number in a third column of Actual Results.

With rare exceptions, all project teams and QA professionals maintain a Defect Log. Even when a project documents nothing else for testing, it maintains a Defect Log and conducts defect management. Teams use a wide variety of tools for managing defects. Some testing teams (employees who contribute to any test asset) use a repository as ordinary as a spreadsheet. Some teams use a tool designated for storage of test assets. These tools explicitly map test assets to upstream assets such as Future State Process Flows (FSPF), User Interface (UI), and Test Conditions, forming Traceability Matrices.

Even though dozens of defect management tools exist, the information captured for individual defects has little variety. A Defect Log assigns an ID number to every defect. A tester adding to the log enters a defect title. They enter the Test Cycle (recorded on every Test Script) and the script step they were executing.

#	Title	Test Cycle	Script Step	Severity	Steps to Re-create Defect	Tester	Status	Assigned To

Testers then assign a severity rating—a universal field in Defect Logs and a common point of contention among testing individuals, project team members, and even the project sponsor. The common terms to prioritize fixing defects are "high, "medium," and "low." Criteria for these include whether testing can continue, the number of customers impacted, and whether a workaround exists for the customer to accomplish what they want (even if only through convoluted action).

Next, testers detail the steps they took to encounter the defect so the employees who contributed to upstream assets can more easily re-create and troubleshoot the defect. The defect tool captures the tester's name, or the tester adds their name manually. The tester sets the status of the defect, such as "new." Finally, the log captures an employee assignment—someone currently working to resolve the defect.

Besides severity, teams also commonly debate whether some defects qualify as enhancements when the tester experiences something upstream assets didn't anticipate, address, or govern. This debate invented the neutral, emotionally uncharged term *system investigation request* (SIR).

Some testing teams manage the Defect Log or "SIR Log" hands-on. An intense test phase might cause the test team to meet every day. Some defect management tools manage workflow, allowing teams to feel hands-off and meet less frequently. In either case, the test team shepherds the right team members to review, revise, and approve the relevant assets so processes and technology can be retested. During the troubleshooting and fixing work, a SIR might reflect a status description of "open," "assigned to fix," "fixed," "assigned to retest," "retested," and "closed." A log should support a status of "rejected" for SIRs a tester recommends for investigation but that the team disagrees about and overrules on. A log should support "deferred" status for unimportant time-consuming fixes and enhancements.

During intense test phases, test teams have daily meetings to discuss the progress of dozens of defects. Managing dozens of open defects resembles a defect factory, and the meetings should manage defects like an assembly line. The first agenda item involves defects retested and ready to change status to "closed." The second agenda item involves fixed defects ready to assign for a retest. The third agenda item involves new and open defects ready to be assigned to fix. Other defect transitions, for example, changing status to fixed or retested, occur outside the test status meeting.

Fixing a defect found in Unit Test is typically simple and quick work because the number of assets to review, revise, and approve is low. But fixing a defect found in a later testing phase (such as a User Acceptance Test) might require revisiting up to a half dozen assets. Even if these changes are straightforward, they might require weeks or months to fix. The magnitude of this negative surprise on a project schedule is the motivation for so much rigor and transparency in upstream assets.

One motivational tool a test team uses to track testing progress (usually at the User Acceptance Test phase) is a graph showing defect quantities during test execution. These graphs have predictable patterns. A graph of *cumulative* defects resembles the S curve of the Actual Results graph. A graph of *open* defects resembles a bell curve (see Figure 12). Inspection of *both* graphs helps determine when a test phase is done or good enough to move on. The first criterion for declaring testing GETMO is that the number of critical open defects is zero. The second criterion is that the number of noncritical open defects is low (in the opinion of the project sponsor). The third criterion is that the Actual Results asset shows a plateauing pass rate above 95 percent (or whatever the project sponsor decides).

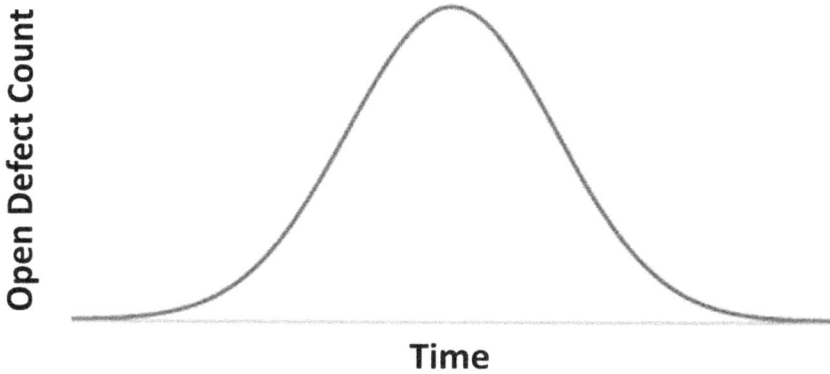

Figure 12. The quantity of open defects typically follows the path of a bell curve.

Five Verbs applies to individual defects in the Defect Log asset—not to the asset in its entirety. Testers *draft* individual items in the asset. As fixing progresses, any testing team member *reviews* and *revises* information about a defect. Every test team meeting aligns on—effectively *approving*—the Defect Log for that day. If the Defect Log is not a constantly available tool, a junior QA analyst *distributes* the log to the test team daily.

Upstream assets of the Defect Log are Test Scripts and Test Data. Its concurrent asset is the test phase's Actual Results. Downstream assets are the Defect Log for other test phases and a Go-Live Announcement.

CONCLUSION

All good men and women must take responsibility to create legacies that will take the next generation to a level we could only imagine.

~ Jim Rohn (1930–2009), American entrepreneur, author, and motivational speaker

Some innovation professionals and their stakeholders have an aversion to transparency and detail. But ambiguity sets others up for confusion, mistakes, and failure. Innovation work benefits from getting "into the weeds," and when teams excel at it, business operations are easier to automate, and business innovation is easier to make automatic.

Healthy innovation teams make the asset portfolio automatic. Instead of creativity in naming the assets and completing the labor (Five Verbs) applied to them, creativity exists within a small number of assets that emphasize the company's market, customers, and stakeholders. The most profitable creativity pertains to these human experiences.

Three groups of project-independent assets capture these human experiences—lens of the market, lens of the team, and lens of the individual. Projects benefit from having current state assets in place before teams rewrite customer experiences in project-specific future state assets. Projects require dozens of assets that can be grouped as process, people, and technology assets. Project management assets provide transparency in, synchronization of, and traceability among the assets.

The number of assets—almost sixty—in the Innovation Elegance methodology might seem intimidating, but most innovation professionals contribute to only a fraction of them. Decentralizing the work minimizes intimidation and boosts confidence in and momentum of teamwork. Each asset is modest and manageable, so disciplined team members can focus their contributions. Disciplined innovation leaders are literate across the entire asset portfolio and shepherd teams to build the assets.

Businesses will find reasons to partition or add to the asset portfolio described in this book. Any collaboration that withstands the rigor of Five Verbs is valuable to the business and durable for future teams. Just as previous methodologies were not static, the Elegance methodology will also evolve. With straightforward, scalable evolution, the asset portfolio of the future can grow to meet the needs of any organization.

Creating documentation is unpopular because it's a lot of work. But benefiting from documentation creation is popular because these assets ease work. The documentation is durable and scalable. It fosters competition (of ideas) and collaboration (of people). It's not more work. It's the real work.

Appendix A: Artificial Intelligence Contributing to the Asset Portfolio

Artificial intelligence (AI) tools such as ChatGPT are increasingly popular. The mainstream hype says that these tools are revolutionizing how teams and businesses work. A more conservative perspective says that a tool like ChatGPT is valuable for *individual* work, but at the moment, its impact on innovation discipline, empathy, and *teamwork* is limited. Also, its correlation with the metaphors of the factory, the asset portfolio, and the empathetic arts is limited. AI's ability to give you insights unique to your customers is doubtful. Its ability to provide you with insights into your team's health is nonexistent.

It's easy to see how AI like ChatGPT can relate to the metaphor of a factory in areas such as *speed*, *automation,* and *ease.* The tool contributes to culture's discipline by educating *individuals*. It acts with empathy because it *listens*, *balances*, and *diversifies* contributions. It is *vigilant*, accepts *rejection,* and behaves with *resilience*. It is an amazing *improvisational* partner, wildly skilled at a collaborative game of "*Yes, And.*" It can contribute to an asset portfolio after it's been educated about the structure of various assets. It seems best suited to contribute to *technology* assets and, among Five Verbs, it seems best suited to *draft* particular assets.

Although it's not a stretch to see how AI improves an individual's discipline, it's difficult to see how it directly improves *team* discipline. It doesn't impact a team's attitudes and behaviors toward competition and collaboration, and thus doesn't shape a culture of team empathy. The tool is skilled at rewriting and summarizing, but its tools are bound to lose detail and to *create* defects. It's unnatural for a project team to assign verbs like *review*, *revise*, and *approve* to ChatGPT. It's also difficult to see how the tool might contribute to project-independent assets, future state processes, or project management assets.

Your team collaborates at or near its unique information frontier. Unless you give an AI tool your complete asset portfolio, it gives you content at a *generic* frontier from the *past*. It's conceivable that ChatGPT contributions are customer-oriented, but it's impossible that those customers are yours. Your innovation teamwork is original, and ChatGPT's contributions are unoriginal, even if they're new to *you*.

Team collaboration requires traceability to upstream assets. Your team has access to upstream assets as well as tribal knowledge not contained in the assets. You'd have to feed the AI tool upstream assets to maintain traceability, but tribal knowledge would be lost. Traceability is a prerequisite for an AI tool to *draft*, *review*, *revise*, or *approve* with legitimacy. For the time being, setting up an AI tool to contribute effectively is laborious, that is, it's a high upfront cost with an unexplored high or low marginal cost.

ChatGPT's strength lies in public information, but what makes innovation difficult is not information but people—how we interact, govern, compete, and collaborate. Humans share private information such as status, lessons learned, and approachability coaching with their coworkers. Humans share what they like, wish, hope, and wonder. For now, AI is not going to contribute to that information. Such tools can directly contribute to *working software* but cannot yet improve the culture of a *working team*.

Although limitations exist, AI tools will continue to improve. It's fascinating to think about what humans will delegate to AI to improve team discipline and empathy. In the past, humans tried to delegate the eight Ds to technology, that is, work that is dull, dirty, dangerous, difficult, demanding, demeaning, delicate, and dear. The economics of innovation don't ask humans to delegate a ninth D—work that is *delightful*. AI will attain a new phase of value when it improves team discipline, empathy, and delight.

Appendix B: Counterproductive Documentation

S ome documents are counterproductive. Certain common and even popular documents undermine discipline and empathy. This appendix identifies ten such documents, explains why they are counterproductive, and specifies what a team should document instead.

These ineffective documents shape cultures of low discipline that promote ambiguity, "reinventing the wheel," and just plain laziness. With low empathy, these documents prioritize data and technology over people. Even the so-called people-centric documents lean toward being self-centric rather than customer-centric, emphasizing content over context.

Documents that are unproductive can feel necessary, but their content, structure, and title can disincentivize discipline and empathy in an innovation team's culture. The following ten documents distract from more valuable documentation. They're ineffective for innovation work and should be avoided.

Data Strategy

Some organizations document a data strategy. A data strategy sets expectations for gathering, storing, curating, and using data to aid decision-making and serve customers more profitably. The goal of a data strategy seems intelligent, but inevitably the document steers a company to be more data-centric, less customer-centric, and less employee-centric. Data centricity emphasizes content management over context management and focuses on *data* context more than on *people* context. A data strategy diminishes the customer experience and the employee experience.

Culturally, a data strategy undermines discipline because the document promotes the idea that organizational *vigilance* relates to information, not people. *Variability* in the actions of customers, employees, and stakeholders is a low priority. A data strategy undermines empathy because it doesn't set expectations for how humans *compete, collaborate,* or *pace* themselves. The document prioritizes information over people.

Data is a wonderful servant but a horrible master. Instead of a data strategy, build and maintain a Customer Experience Hierarchy. It shapes a culture of discipline and empathy for humans—customers, employees, and extended stakeholders.

Technology Roadmap

Some organizations document a technology roadmap that sets expectations for hardware and software investments. The typical goals of a technology roadmap are to reduce the costs of technology or labor. This document seems like a good idea, but it undermines the customer-centric Roadmap and hurts customer centricity. Likely, sales teams and customer-facing employees are minimally involved in a technology roadmap, so the document is prone to neglecting revenue and profit.

When technology actors perform poorly or approach obsolescence, re-engineer or replace them as part of customer-centric innovation in a customer-centric Roadmap. The Innovation Elegance Roadmap keeps revenue, cost, and profit in mind.

Culturally, a technology roadmap undermines discipline because it allows team members to lose sight of the *economics* of customer centricity and the *quality* of nontechnology teamwork. The document undermines empathy because it is *attentive* to the needs of technology actors instead of people actors.

Technology is a wonderful servant but a horrible master. Instead of a technology roadmap, build and maintain a people-centric Roadmap. It shapes a culture of discipline and empathy for humans—customers, employees, and extended stakeholders.

Meeting Minutes

Many organizations document meeting minutes. Meeting minutes capture (in writing) decisions, next steps, and general conversation highlights. Creating this written record seems like a responsible thing to do after any meeting, because any document is better than no document, right?

But meeting minutes have disadvantages. The document reveals that the team lacked a predefined structure for the collaboration. The lack of structure neglects core expectation setting, dilutes attention to necessary alignment, and makes the meeting vulnerable to wandering into low-priority topics. Such a document is irresponsibly tolerant of strong personalities monopolizing meetings and tolerant of attendees not knowing how sincerely aligned they are. Conventional wisdom is that the value of meeting minutes quickly depreciates. Meeting minutes are disposable (especially when what success means for a meeting is that it has meeting minutes).

Conversations are most valuable when the team knows in advance the asset to discuss—its structure, boundaries, and scope. Healthy teams grasp that the point of the meeting is to review, revise, and approve the asset. Unstructured documentation, such as meeting minutes, evades accountability and is difficult to improve incrementally. In contrast, structured documentation is conducive to small improvements, collaboration, approval, and accountability.

Culturally, meeting minutes undermine discipline because they are vulnerable to high *variability* (i.e., being "all over the place"). They *waste* people's time because people should be collaborating on high-quality, highly durable documentation instead. The document undermines empathy because employees lack *trust* in their collaboration. Meeting minutes promote a culture of volatility, uncertainty, complexity, and ambiguity (VUCA) because meetings are often volatile, uncertain, complex, and ambiguous when structured documentation is not the focus.

Instead of meeting minutes, formalize the structured assets—your asset portfolio—that justify meetings. *Draft* the structured asset before the meeting. *Review* and *revise* it in the meeting—where ideas can compete, and people can collaborate. *Approve* and *distribute* the asset. Five Verbs minimizes VUCA and promote a culture of discipline and empathy.

Team Charter

Some teams document a team charter. This document resembles a Project Charter because it includes objectives, scope, and names of people. Like a Project Charter, a team charter aims to reduce ambiguity.

A team charter is appropriate for a team's sustainable operations, but not for a team's innovation work. Healthy innovation is customer-centric, and a team charter is self-centric.

During any project, assignments can shift. A Project Plan easily accommodates shifts in dependencies, duration, and assignments, whereas a team charter contains only some of this information. Assignments in a team charter are typically high level, and ambiguity and duplicates are common, undermining clear assignments such as those in a Project Plan.

One fundamental expectation of any kind of charter is that its contents are stable for its scope and schedule. When a team charter experiences a change, it can feel like the team is changing the rules in the middle of the game. However, revising a Project Charter shows attentiveness and transparency. Innovation projects experience change; being governed by a team charter instead of a Project Charter discourages attentiveness and suppresses transparency.

Culturally, a team charter undermines discipline because its *accountability* is to itself instead of to a customer. A team charter undermines empathy because it *discourages revisions* and cultivates self-centricity at the expense of customer centricity. Instead of a team charter, document a Project Charter.

RACI Matrix

Many teams create a RACI matrix, which stands for responsible, accountable, consulted, and informed. A RACI matrix lists work items on one axis, lists roles or team members' names on the other axis, and assigns a letter—*R*, *A*, *C*, or *I*—at the intersections to establish someone's level of involvement.

Assignments are always helpful, but a RACI matrix harbors a lot of ambiguity. It distracts from the Project Plan, which accomplishes everything a RACI matrix wants to, but with less ambiguity, more information, and more *durability*. In reality, RACI matrices don't create the intended clarity and discipline but only divert attention from a Project Plan, which governs with clarity and discipline.

Culturally, a RACI matrix undermines discipline because the word choice in the matrix doesn't demand documentation and the *transparency* it provides. A RACI matrix also provides a false sense of discipline, which undermines empathy because it preserves *ambiguity* about the sequence of work, estimated duration, and progress toward completion. Instead of a RACI matrix's four letters, focus on Five Verbs (draft, review, revise, approve, distribute) in a Project Plan.

RAIL and Task List

Some teams maintain a RAIL—a "rolling action items log." A RAIL usually resides in a spreadsheet; it assigns and prioritizes work and tracks progress toward completion. Similarly, many teams maintain a task list, which is similar in intent, but which omits priority and progress.

Although it's better to have a single written form of a team's work than no written form, these documents reveal two massive points of neglect. First, a team with either a task list or a RAIL is envisioning its work only one to two weeks into the future and no further, thus reinventing the wheel with its work. Second, such a team is building a new road as they drive on it, whereas Elegant innovation teams have already built the road. At their worst, these two documents are public dumping grounds of impulsively conceived laboriousness.

If you trim away wasteful and noisy work, practically all work on healthy projects is the same. Every project has a finite and manageable quantity of documents to build. Managing the infinite number of meetings and emails necessary to build a task list or a RAIL is unrealistic and counterproductive. But a Project Plan with Five Verbs is realistic, with no waste or noise.

Culturally, task lists and RAILs undermine discipline because *reinventing the wheel* hurts a team's *speed* in planning and executing a project. Language in these documents is *highly variable*—reliably all over the place with Verb Sprawl. These documents also undermine empathy because their content exhibits VUCA (volatility, uncertainty, complexity, and ambiguity) and moves the team away from *visible*, thoughtfully sequenced, *paced*, collaborative documentation.

Instead of a task list or a RAIL, build and maintain a Project Plan that includes every document relevant to innovation work, governs them with Five Verbs, and nothing else.

Business Requirements and Rules

Many project teams document business rules or business requirements. By their name, these documents aim to be repositories for what "business" employees want from their technology. Near the beginning of an innovation project, it's better to have business-oriented documentation instead of technology-oriented or no documentation. However, these documents typically lack the valuable structure and context for small improvements, small projects, and rich customer experience. At their worst, they are a list of instructions for a developer or Software as a Service (SaaS) administrator—design specifications written by a "businessperson."

References to "business" arouse a distinction between business and technology that is no longer useful and can even be counterproductive. Whatever a team might envision including in these documents should focus on customers and employees and their context, story, and journey. The proper asset is Future State Process Flows (FSPF).

Business requirements and rules imply a detachment from current state processes and a disregard about what is dysfunctional about the current state of customer and employee experiences. The terms *requirements* and *rules* imply self-centered, egotistical demands by a few employees. Innovation is never about yourself. Innovation is always grounded in your customer and in the employees closer to revenue.

Culturally, business requirements and rules undermine discipline because they dilute attention on *process governance* and the *context* of the impacted stakeholders. These documents undermine empathy because they advocate for an *ambiguous* entity (i.e., "someone other than IT"). Their label "bangs the table" without customer or employee context and story. Instead of these documents, build FSPF with Current State Process Flows (CSPF) already in hand.

FAQ

Many companies generate Frequently Asked Questions (FAQ). An FAQ diligently gathers the most common or expected questions a team wants to answer and gives the explanations to stakeholders. The intent is to show responsiveness, clarity, and transparency and to reduce communication traffic.

But an FAQ separates information from the assets where it truly belongs. An FAQ formalizes that upstream assets lack completeness, accuracy, or precision. An FAQ broadcasts, "Look at all our ambiguous, low-quality work that customers found instead of us!" An FAQ also suggests that the project team proposed a workaround instead of fixing the problem at its origin. It hints that the team will tolerate the problems—defects—in the upstream assets. Instead, the team should revise those assets.

One asset to review and revise is training materials. If revisions are limited to training materials, impacts on the project schedule, rework, and value proposition are likely modest. The other asset an FAQ can expose problems with is Future State Process Flows (FSPF). Problems in process flows—exposed only when a team publishes an FAQ—can have severe impacts, including rework, schedule delays, and even a jeopardized value proposition.

Culturally, an FAQ undermines discipline because it shrugs at the low *quality* of upstream assets. An FAQ undermines empathy because it creates *messiness* and *laboriousness* for customers and procrastinates on fixing upstream assets for future project teams. Instead of publishing an FAQ, improve quality at the origins of problems by reviewing, revising, approving, and distributing the (previously) problematic assets.

Appendix C: Detailed Project Planning

The Elegance methodology contains approximately sixty assets and proposes a five-verb framework to govern the assets. Projects vary by name, assignments, and duration, of course, but the relevant assets and necessary verbs are very stable. Projects should not reinvent the wheel. When you know the relationships of the assets, building a default Project Plan is methodical. This appendix presents a default Project Plan for any team to tailor for its own assignments and duration estimates.

Starting from a blank Project Plan containing only a project start date (January 10 in the examples below), list assets and populate Five Verbs on four rows (*review* and *revise* can share a row). Populate dependencies (row numbers) so the row sequence is Draft, Review and Revise, Approve, and Distribute. Assign each row to the appropriate team members. Assuming your project management tool defaults duration estimates to one day, consult team members on whether they wish to reserve more days and let the detail reflect the number they want to reserve. The typical project management tool derives dates from dependencies and durations. Every week, update the percent-complete field based on the information you receive from team members, that is, their Individual Status Reports.

The following sample asset portfolio is organized in three sections: project-independent, project-specific, and technology assets.

Project-Independent Assets

The asset portfolio contains approximately twenty project-independent assets. A first pass at creating the assets follows the sequence of how they appear in this book. Ongoing revisions follow a frequency (quarterly, monthly, or weekly) and can be staggered to avoid work spikes and to encourage the cross-impact that assets have on each other.

Assets revised quarterly can be staggered *monthly* as shown in Figure 13.

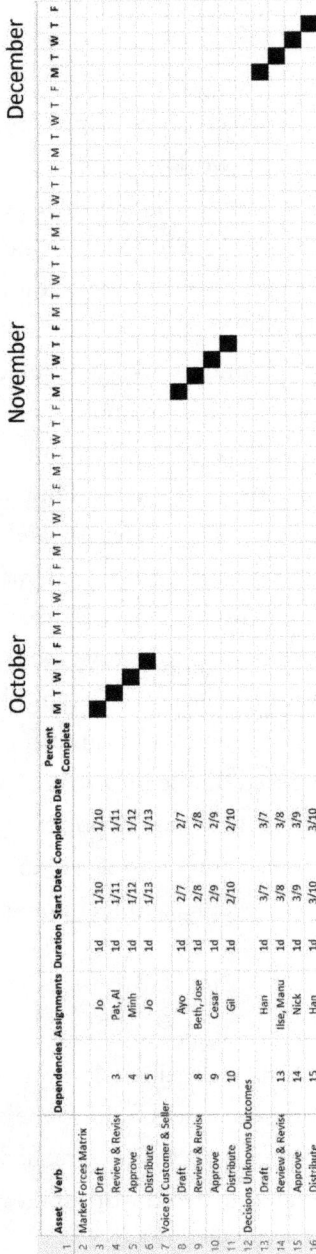

Asset	Verb	Dependencies	Assignments	Duration	Start Date	Completion Date	Percent Complete
Market Forces Matrix							
	Draft		Jo	1d	1/10	1/10	
	Review & Revise	3	Pat, Al	1d	1/11	1/11	
	Approve	4	Minh	1d	1/12	1/12	
	Distribute	5	Jo	1d	1/13	1/13	
Voice of Customer & Seller							
	Draft		Ayo	1d	2/7	2/7	
	Review & Revise	8	Beth, Jose	1d	2/8	2/8	
	Approve	9	Cesar	1d	2/9	2/9	
	Distribute	10	Gil	1d	2/10	2/10	
Decisions Unknowns Outcomes							
	Draft		Han	1d	3/7	3/7	
	Review & Revise	13	Ilse, Manu	1d	3/8	3/8	
	Approve	14	Nick	1d	3/9	3/9	
	Distribute	15	Han	1d	3/10	3/10	

Figure 13. Quarterly assets staggered monthly.

Assets revised monthly can be staggered *weekly* as shown in Figure 14.

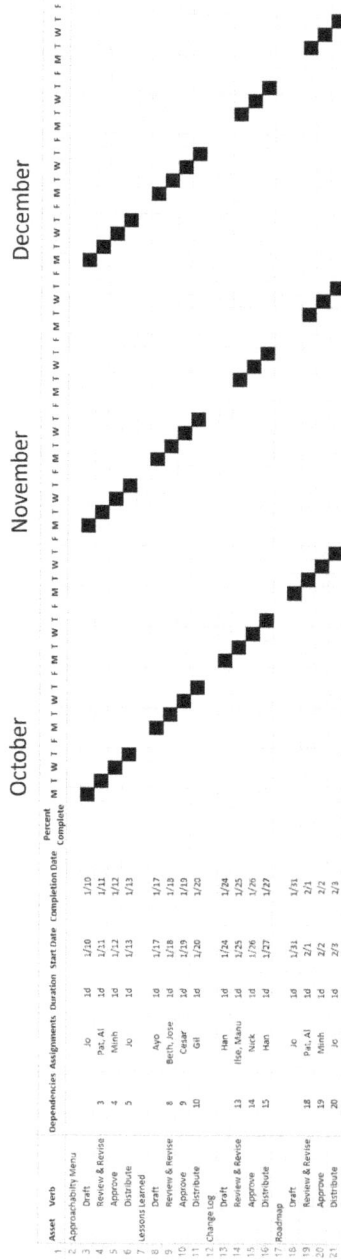

Figure 14. Monthly assets staggered weekly

Assets at a weekly rhythm might resemble Figure 15.

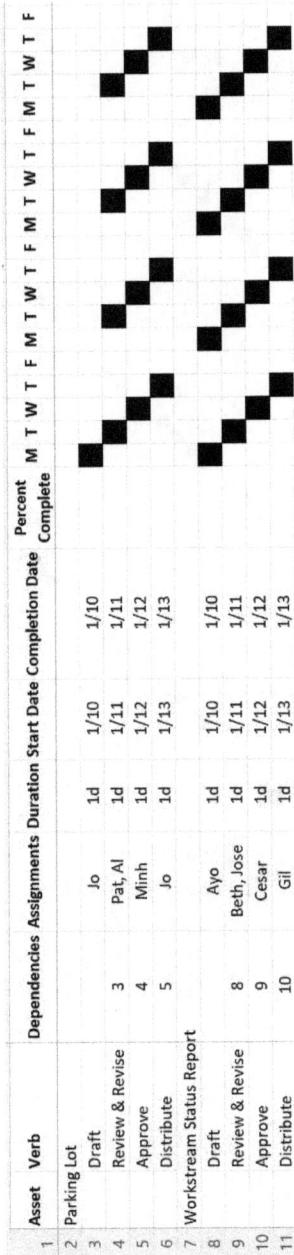

	Asset	Verb	Dependencies	Assignments	Duration	Start Date	Completion Date	Percent Complete
1	Parking Lot							
2		Draft		Jo	1d	1/10	1/10	
3		Review & Revise	3	Pat, Al	1d	1/11	1/11	
4		Approve	4	Minh	1d	1/12	1/12	
5		Distribute	5	Jo	1d	1/13	1/13	
6								
7	Workstream Status Report							
8		Draft		Ayo	1d	1/10	1/10	
9		Review & Revise	8	Beth, Jose	1d	1/11	1/11	
10		Approve	9	Cesar	1d	1/12	1/12	
11		Distribute	10	Gil	1d	1/13	1/13	

Figure 15. Weekly assets.

Creating current state assets represents getting out of documentation debt. Like any debt, paying it back too quickly or slowly is painful. Creating current state assets too quickly causes projects to wait until team members are available. Creating current state assets too slowly causes projects to start in debt. You can plan and pace paying back documentation debt. One pass, for example, to set up one project to start debt-free, might look like this five-week payback plan shown in Figure 16.

There are a couple of exceptions to planning and pacing project-independent assets. A Scorecard likely contains metrics measured at any of the frequencies illustrated above (quarterly, monthly, or weekly). It's complicated to reflect this variability in a Project Plan. For lens of the individual assets—such as Individual Status Reports and Workload Reports—employees draft them (weekly), managers review them (weekly), and there's no need to revise, approve, and distribute them. A Project Plan is overkill for assets with so few contributors. All this said, when project-independent assets can benefit from transparency of assignments, pacing, and a rising percentage completion, a project planning tool improves discipline and empathy in the work.

Project-Specific and Technology Assets

For an individual project, weekly and monthly *repetitions* don't apply. Instead, plan with *sequence* in mind and *synchronize* process, people, and technology assets. Figure 17 is a default Project Plan that reflects one day each for draft, review and revise, approve, and distribute. The first three weeks show daily granularity. For legibility, weeks four through twelve collapse Five Verbs into a single block of work and show weekly granularity. The plan contains four test phases and proposes a project duration of twelve weeks.

In your project planning tool, populate your Project Plan top to bottom and left to right. First, build the asset/verb hierarchy. Second, add dependencies (similar to the examples above but shown here only visually). Next, assign asset/verb combinations to team members (like the examples above). Next, consult the assigned team members on their duration esti-

Asset	Verb	Dependencies	Assignments	Duration	Start Date	Completion Date	Percent Complete	M T W T F M T W T F M T W T F M T W T F M T W T F
Current State Inventory								
	Draft		Jo	1d	1/10	1/10		
	Review & Revise	3	Pat, Al	1d	1/11	1/11		
	Approve	4	Minh	1d	1/12	1/12		
	Distribute	5	Jo	1d	1/13	1/13		
Current State Scripts								
	Draft	6	Ayo	1d	1/17	1/17		
	Review & Revise	8	Beth, Jose	1d	1/18	1/18		
	Approve	9	Cesar	1d	1/19	1/19		
	Distribute	10	Gil	1d	1/20	1/20		
Current State Process Flows								
	Draft	11	Han	1d	1/24	1/24		
	Review & Revise	13	Ilse, Manu	1d	1/25	1/25		
	Approve	14	Nick	1d	1/26	1/26		
	Distribute	15	Han	1d	1/27	1/27		
Customer Experience Hierarchy								
	Draft	16	Jo	1d	1/31	1/31		
	Review & Revise	18	Pat, Al	1d	2/1	2/1		
	Approve	19	Minh	1d	2/2	2/2		
	Distribute	20	Jo	1d	2/3	2/3		
Use Case Assessment								
	Draft	21	Jo	1d	2/7	2/7		
	Review & Revise	23	Pat, Al	1d	2/8	2/8		
	Approve	24	Minh	1d	2/9	2/9		
	Distribute	25	Jo	1d	2/10	2/10		

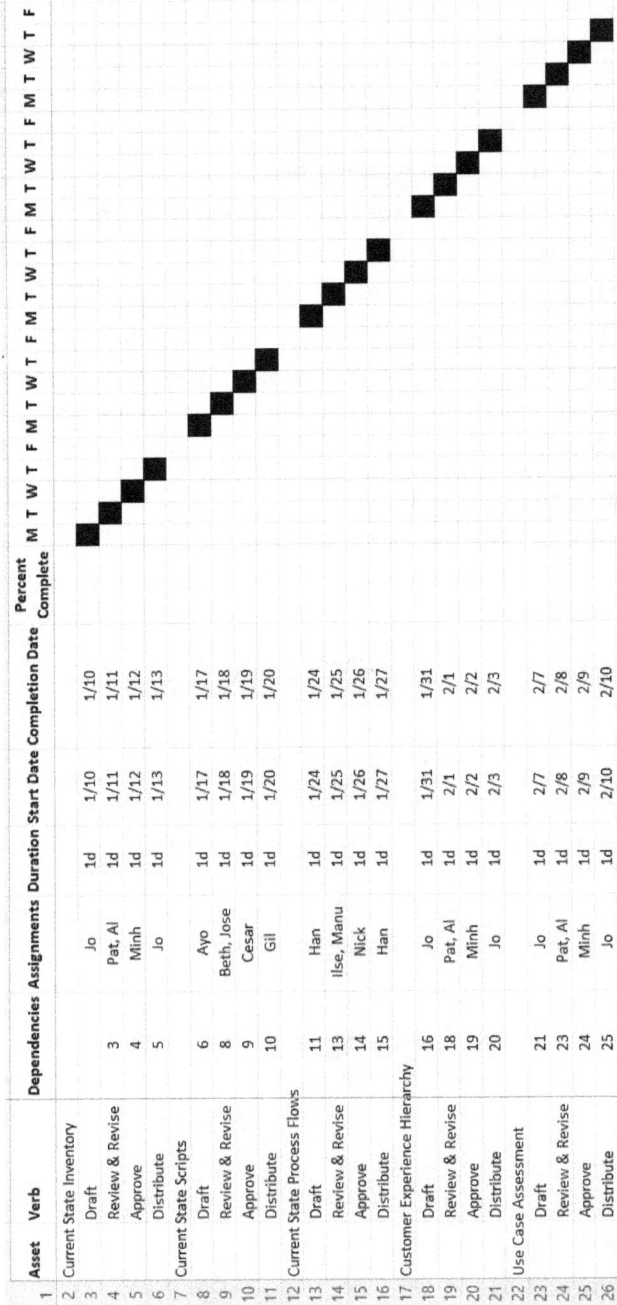

Figure 16. Current State assets.

mate (some will want to reserve more than one day). Finally, as the project progresses, update (approximately weekly) the percent complete values (placeholders in examples above).

Go-live events using software-centric methodologies are often too close or too far apart. The Elegance methodology aims for a twelve-week synchronized, visible, and elastic rhythm. Instead of VUCA (volatility, uncertainty, complexity, and ambiguity), innovation work is simple, stable, confident, and clear—ready for new information anytime so that it can pause, pivot, and proceed.

When a team feels skilled in synchronization, it might be comfortable partially overlapping projects and starting them more frequently than every twelve weeks. A team could start new projects six weeks, four weeks, or even three weeks apart. Staggering projects this way resembles performing music in a round—imitating songs such as *Frère Jacques* and *Row, Row, Row Your Boat* that stagger their melody, forming harmony. See the next page for an illustration.

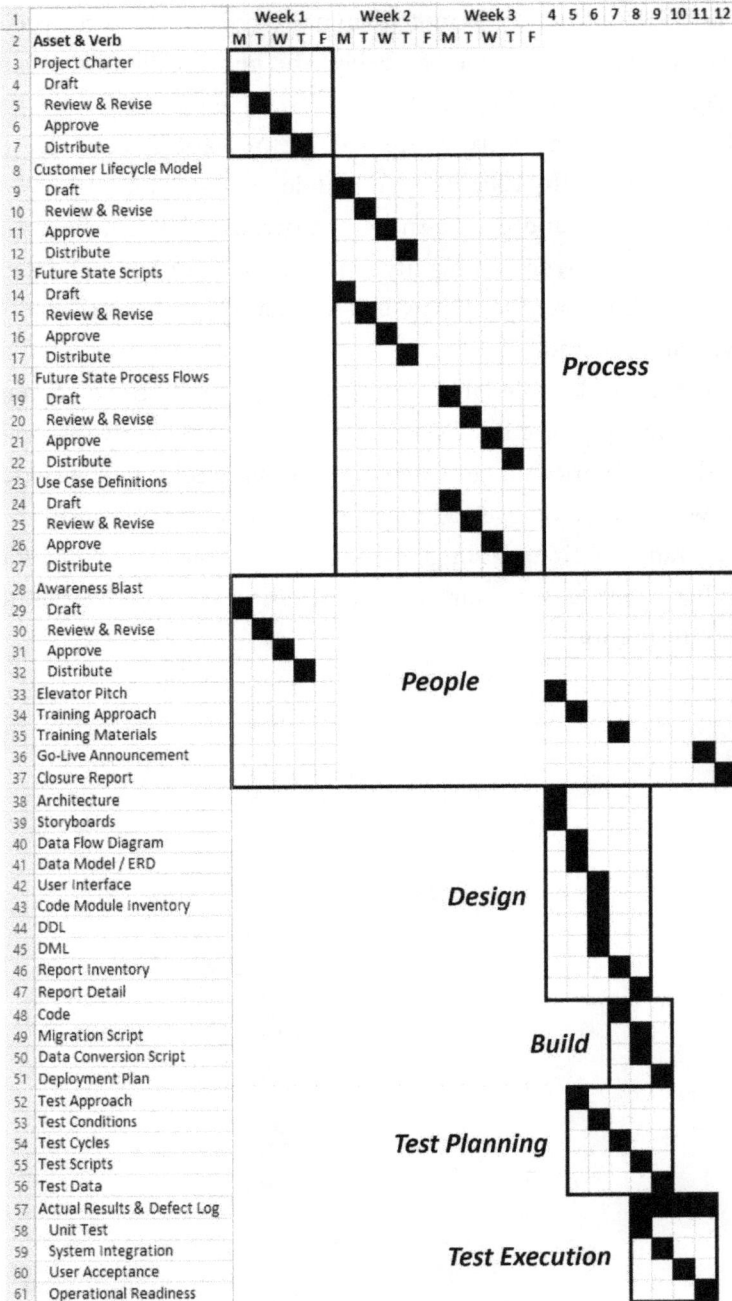

Figure 17. An example of a twelve-week project.

Appendix D: Lessons Learned Deep Dive

The Lessons Learned template has fourteen topics for a team to reflect on. Some innovation professionals feel they grasp each topic immediately upon reading it. But some topics warrant more description than the title alone. This appendix shares descriptions.

Like the template itself, these topics are ordered by urgency. A team new to such a rigorous group reflection starts with the first few (three or four) topics. As a team gains familiarity with the template and reflections become more positive, it can adopt more. Adopting them gradually enables a team to increase discipline and empathy in their culture simultaneously at a comfortable pace.

Safety, Inclusivity, and Belonging

A team has healthy safety, inclusivity, and belonging (SIB) when every person works with minimal fear for themselves. They may not control their job security, but they have faith in their career security. Because they don't fret about themselves, they unleash their time, attention, and passion for their work.

Without SIB, employees become self-absorbed and self-centric. Instead of collaborating, they compete and feel incentivized to reduce others' value and safety. When employees feel safe, included, and like they belong, they are optimistic about their success and free to be ambitious for themselves, their colleagues, and customers.

Three assets are instrumental for fostering a culture of SIB. The Project Plan shows who has a "seat at the table." I Like I Wish I Hope I Wonder welcomes weekly positive and negative reflections. The Approachability Menu encourages employees to optimize their contributions across meetings, emails, and assets and to mentor others to do the same.

Transparency

Transparency in innovation work takes a few different forms. One form shows the assignments, timing, and contents of assets. Another form knows the most problematic parts of the current customer experience that warrant innovation. A third form shares assessments of project health.

Poor transparency breeds poor prioritization, rework, and negative surprises. Adding laboriousness and extending schedules hurts the value proposition for all project stakeholders. Transparency is vital for managing expectations and perceptions of innovation teamwork. Healthy transparency helps innovation speed, quality, and ease.

Every asset in the asset portfolio eases information sharing and contributes to a culture of transparency. Assets that boost transparency include the Awareness Blast, which engages low-profile stakeholders. The Use Case Assessment exposes the most problematic parts of every customer experience. The Stoplight Report acknowledges whether a project's health is poor, at risk, or good.

Simple and Straightforward

A team has healthy simplicity when it avoids sprawl. To keep the customer experience and employee experience simple and straightforward, nothing should feel infinite. The size or quantity of anything must be small enough to provide a sense of progress and moving on.

The opposite of simplicity is complexity. Some innovation stakeholders have the incentive to make work complex or make work seem complex. In teamwork, complexity is a silent killer. Most often, bad complexity survives in meetings and emails that shun the asset portfolio. Bad complexity thrives in exotic word choices and Verb Sprawl in Project Plans. The Five Verbs framework is a reaction to Verb Sprawl to enforce simplicity on an innovation team.

The asset most instrumental to *innovation* simplicity is the Project Plan. The finite quantities of assets and verbs instill confidence and clarity in an innovation team. The assets most instrumental to *operational* simplicity are process flows. Swimlanes, boxes, and arrows set expectations for who does what and when. These assets challenge complexity, encourage assignments that are straightforward, and have a domino effect that fosters simplicity in other assets.

Accountability

A team has healthy accountability when projects and individuals do what they say they will do—aka a "say/do ratio" equal to one. Overpromising is common. Mature innovation teams know that poor accountability is a constant risk.

One lens of accountability is through trust. A project has healthy accountability when team members have competence, integrity, honesty, and benevolence to fulfill their assignments. Poor accountability results in low quality, ethical problems, and relationship breakdowns. Healthy accountability results in responsiveness, speed, and high-quality teamwork.

Assets instrumental to a single employee's say/do ratio are Project Plans (say what each person will do) and Individual Status Reports (say what each person did). Assets instrumental to a project's say/do ratio are the Project Charter (say what the project will do) and the Closure Report (say what the project accomplished).

Alignment

A team has broad, healthy alignment when it has 1001 small alignments. Alignment is critical for a team to have a sense of shared goals, values, and purpose.

Poor alignment originates in diverging priorities, poor listening, and low empathy. These culture traits make teamwork pointless. Poorly aligned employees should not work together. Instead, employees who constantly keep a bigger picture in mind care about colleagues, listen to them, and compromise on priorities.

The assets most instrumental to alignment are the Workstream Status Report and Individual Status Report. These represent unified statements of reality—shared and reinforced weekly. Of the Five Verbs, *approve* is vital to convert diverging ideas into converging ideas, to resolve disagreements, and to proceed with a unified vision of purpose, goals, and action.

Momentum

Activity within a project team building an asset portfolio resembles a factory. The team repeats Five Verbs for every project-specific asset. A team has healthy momentum when it approves assets at a reasonable and legitimate pace—not too fast or slow.

Momentum is a problem when a team has the wrong organizational friction. Too much friction is a sign of perfectionism, bureaucracy, or the lack of a tiebreaker. Too little friction is siloed, exclusionary work that lacks a bigger picture and causes rework. "Goldilocks"—just the right amount of—friction has sufficient contributors, preassigned tiebreakers to formalize every approval, and the maturity to declare a small win with every instance of reaching GETMO.

Three assets govern team momentum. Status reports reveal the weekly pace of asset approvals. The Project Plan records the percent complete for each of the Five Verbs as every station of the asset factory, at a reasonable rate, reaches 100 percent. Go-Live Announcements show that teams finish what they start.

Morale

Employees don't have to be empathetic to know that morale impacts the ability to collaborate, the value of teamwork, and ultimately, profit.

A team with positive morale has stability, confidence, and clarity that keep its performance high.

Attitude and morale are easily confused, but they're different. An employer cannot dictate attitude, but it certainly shapes morale. When a team culture provides safety and a sense of belonging, inspires confidence in the work, and feels like an excellent employment choice, morale will be high. Enthusiasm breeds impressive speed and quality in launching improved customer experiences.

Methodology is the leading feature of culture. Morale is a trailing feature—at the mercy of every other topic in this Lessons Learned template. Even so, morale is a form of feedback that mentors the methodology in return. Leaders shouldn't pressure employees for enthusiasm but should use a tool and maintain a culture that cares about morale. The asset that captures morale is I Like I Wish I Hope I Wonder.

Sustainability

Teamwork has sustainability when its risk of collapse is low. Striving for sustainability is a bland goal, but innovation is an ambitious business. Unsustainability is a legitimate, significant, and widespread risk. Ambition is only great when the journey and the benefits are sustainable.

Sustainability in teamwork has one quantitative ingredient—workload—and one qualitative ingredient—hostility. If an employee's workload is too high, they (and their work) are vulnerable to collapse. If they see, experience, or are aware of hostility, they (and their work) are susceptible to collapse. Disciplined, empathetic employees have selfish reasons for proactively monitoring sustainability and minimizing their own vulnerability to a colleague's collapse.

Three assets govern an innovation team's sustainability. A Roadmap reveals the number of simultaneous projects and hints at the workload across dozens of employees. Tolerating too many projects jeopardizes sustainability for countless employees. The Workload Report reveals one employee's sustainability. Tolerating too many work hours threatens sustainability for a single employee. The Approachability Menu tells where meeting and email etiquette must change to eliminate hostility. Tolerating hostility jeopardizes sustainability for exposed employees.

Scalability

A team has scalability when it constantly pursues low marginal cost. For workers, this translates to low laboriousness—activities that don't fatigue. Scalability is instrumental for company profits and so that employees have energy and attentiveness when new information and opportunities appear.

Cost-cutting is a common theme in technology. Poor scalability is common in innovation teams. A team is not scalable when the culture is overwhelmed with meetings and email. These communication channels are a common addiction and a comfort zone because they have low upfront costs. Meeting gridlock and email overload are painful because they have high marginal costs. They are fatiguing, laborious, and not scalable. The fundamental premise of this book is that non-email documentation—governed by Five Verbs—is the communication channel with low marginal cost and low laboriousness. As the pace of innovation increases, low costs and scalability magnify their impact on profitability.

While the asset portfolio is the centerpiece for scalability, one low-profile asset uniquely monitors and mentors scalability—the Pie Chart. It reports on high operational burdens, unplanned work, and chaos—revealing laboriousness and where scalability can be improved.

Another form of scalability is a team improving its skill in applying Five Verbs to synchronize its asset portfolio. For example, a team might be able to reduce the duration of their default projects from twelve weeks to six weeks.

Stylishness

A team's work is stylish if employees perceive that the outside world (friends, family, or past colleagues) sees their work as interesting and cool. Employees take pride in their work and are eager to add highlights to social media and their résumé. Stylishness is not all that common, so it's valuable for a team to recognize and appreciate the moment while it's happening. Even if team members shrug at the work's popularity, buzz about it can create a virtuous circle of speed, quality, and positive surprises.

There's no shame in work that lacks cachet. Some innovation work is kept quiet for competitive, regulatory, or reputational purposes. Many projects avoid glamour because of long work hours or a tense environment. Work can be subdued because being upbeat and playful lacks awareness of customer or employee circumstances. "Putting lipstick on a pig" is tone deaf and insulting. Acknowledging cheerlessness is empathetic and salvages credibility among stakeholders.

At a team level, the asset most instrumental to stylishness is the Elevator Pitch. The Elevator Pitch inspires and reinforces "Why us? Why this? Why now?" At an individual level, the Individual Status Report reflects stylishness and seriousness. Status reports facilitate the story every employee wants to tell about their contribution once the project is done.

Learning

For most team members, all innovation work is educational in some way. Distinctive questions about learning ask whether employees feel they're learning *during* the project, whether they realize learning *after* the project, and whether the learning is what they need or want for their career.

All projects and project teams are different, so consciously or not, employees learn from their team members, customers, and more distant industry players. Employees conclude which work is positive or negative and which people are positive or negative role models. Reflective employees ask themselves, "Do I like this work?" "Are these skills transferable?" "How would I lead or follow differently?" and "How do I want my future work to be different?"

The assets most instrumental to learning are the Project Charter and Project Plan. Team members learn more when there is variety in Project Charters and their assignments. Employees learn less when they contribute to similar projects and assets with the same people, but they learn more when they contribute to diverse projects and assets with different people.

Emphasis

A team has healthy emphasis when the essence of a project stays consistent during the project. Examples of project emphases are revenue or cost orientation, customer centricity, employee centricity or behind-the-scenes work, and serving new or existing processes and people. A rigorous Lessons Learned exercise checks for the team's fidelity to emphasis during the project.

In monthly Lessons Learned exercises, teams watch whether the most current work aligns with what the team expected at the start of the project. If not, the team may have lost sight of the value proposition. Additional examples of project emphasis include qualitative versus quantitative change, high-touch versus low-touch (customer intimacy versus automation), and process governance versus data governance.

Assets that convey the emphasis of a project are the Scorecard, Project Charter, Use Case Definitions, and training materials. Emphasis is a problem when these assets diverge, but emphasis is healthy when comparing these assets concludes they belong to the same project.

Balance

A team has a healthy balance when certain assets contain some variety. The variety comes in the form of projects, mild levels of idle work, and mild levels of idle workers. Even healthy teams are never securely in balance but constantly correct for *imbalance*.

Examples of balance include project objectives having different emphases, all team members consistently and sustainably contributing, and work rarely stalling. Balance includes Workload Reports showing no one is grossly overworking or undercontributing.

A few assets help teams correct imbalance. The Roadmap shows idle work and a variety of work. A Project Plan shows concentrated and dispersed contributions. Individual Status Reports show idle work. Workload Reports show idle workers.

Success Is Inevitable

The inevitability of innovation success sits at a different altitude than the other topics in the Lessons Learned template. It is a catchall. As comprehensive as the previous thirteen topics are for a Lessons Learned exercise, a stakeholder might think of a concern that doesn't fit among these topics. This topic in the template can steer them to articulate their idea.

Even with extraordinary discipline and empathy, innovation success is never guaranteed. No one expects success rates to be perfect. But this is not a license to be complacent and tolerate problems hanging around and turning systemic. Other topics in the Lessons Learned template aim to minimize ambiguity, and the prompt "Success is inevitable" is purposely ambiguous. It resembles a clear blue sky for stakeholders to ask, "What else? Who else? How else?"

The asset unique to striving for inevitable success (after optimizing every other aspect of the Lessons Learned exercise) is the Parking Lot. Amid dozens of assets innovation teams need for success, the Parking Lot acknowledges the barriers, risks, issues, and questions that arise that don't neatly fit into the other assets. The Lessons Learned template labels the systemic and systematic problems in innovation work as the "pattern in the problems." The template sets the table for teams to minimize these patterns. The discipline in the template eases navigating and absorbing sampling errors, that is, human mistakes, with grace and elegance.

For convenience, the table below maps topics in the Lessons Learned asset to assets most instrumental for improving each culture trait.

Culture Trait	Assets
Safety, Inclusivity, Belonging	I Like I Wish I Hope I Wonder; Approachability Menu; Project Plan
Transparency	Use Case Assessment; Awareness Blast; Stoplight Report
Simple and Straightforward	Project Plan; Process Flows
Accountability	Individual Status Reports; Project Charter; Project Plan; Closure Report
Alignment	Workstream Status Report; The verb "Approve"; Individual Status Report
Momentum	Workstream Status Report; Go-Live Announcement; Project Plan
Morale	I Like I Wish I Hope I Wonder
Sustainability	Roadmap; Approachability Menu; Workload Report
Scalability	Executing Five Verbs over time; Pie Chart
Stylishness	Individual Status Report; Elevator Pitch
Learning	Project Charter
Emphasis	Scorecard; Use Case Definitions; Project Charter; Training Materials
Balance	Roadmap; Individual Status Reports; Project Plan; Workload Report
Success Is Inevitable	Parking Lot

Best and Worst-Case Scenarios

Below are two populated Lessons Learned exercises containing ideas and humor for your team. One describes a best-case scenario, and one describes a worst-case scenario.

Worst-Case Scenario

Topic	I like	I wish I hope	I wonder
Safety, Inclusivity, Belonging		I wish I wasn't afraid of my boss. I wish I didn't despise my coworkers. I wish everything wasn't CYA. I wish I could reject their lunch invitation, but I'd have to be invited first.	
Transparency		I wish I knew what was going on with the project. I wish I understood the bigger picture.	
Simple and Straightforward		I wish every meeting didn't have a surprise. I wish the process weren't so complicated and so locked inside a few people's heads.	
Accountability		I wish people knew their job, actually did their job, only their job, or just got another job.	
Alignment		I hope my coworker isn't suppressing their concerns. I hope our other lines of business were consulted.	
Momentum		I hope our work won't always seem too fast (siloed, frantic) or too slow (bureaucracy and interruptions).	
Morale		I wish we didn't have so much nervousness around the team.	
Sustainability		I wish my coworkers weren't always on the edge of burnout. I wish we didn't have so many fire drills.	
Scalability		I wish we weren't so dependent on so few people.	
Stylish		I hope this project will look good on my LinkedIn profile.	
Learning		I hope this work feels applicable to other positions, teams, or companies. I hope I learn something new soon.	
Emphasis		I wish I wasn't always fixing data.	
Balance		I wish I wasn't in meeting gridlock and email hell.	
Success Is Inevitable		I wish I had some confidence in any of our projects, their true value propositions, or our methodology. I hope this project wasn't doomed before it started.	

Best-Case Scenario

Topic	I like	I wish / I hope	I wonder
Safety, Inclusivity, Belonging	I like that no one is shy on this team. I like that the team seems comfortable around one another. I like that my team has very different people eager to work together.		
Transparency	I like our frequent comment, "Our future team will appreciate this documentation!" I like that no one is surprised about their assignments.		
Simple and Straightforward	I like that we scrutinize processes when they're difficult to draw. I like that nothing about the project or operations feels convoluted or burdensome.		
Accountability	I like how visible and clean our Project Plan is. Everyone knows their lane.		
Alignment	I like that we minimize silent dissent. People do speak up! I like that we use our tiebreakers to resolve task conflicts. Personality conflict is low.		
Momentum	I like that we rarely feel too fast or too slow.		
Morale	I like our combination of seriousness and playfulness. Feedback is constructive.		
Sustainability	I like that no one is working themselves toward burnout. We've had no crisis.		
Scalability	I like that we reengineer and automate laborious work.		
Stylish	I like putting this work on my résumé/LinkedIn profile.		
Learning	I like that none of this work feels old, and a lot of it feels new.		
Emphasis	I like that we emphasize customers and employees rather than technology or data. I like that we have attention to detail but aren't overboard.		
Balance	I like that the skills across the team feel like a good fit. I like our balance across meetings, email, and documentation outside email.		
Success Is Inevitable	I like that everyone's confidence and pride in the project are so high. I like that the team feels our project success rate from now on will be really high.		

Afterword

Freedom is ... spending time to imagine that which you cannot yet imagine.
The freedom to imagine that as yet unimaginable work in front of others,
moving them to still more action you yourself cannot imagine.

~ Jim Cohen (b. 1949), American innovation design strategist and speaker

Innovation projects are more difficult than they should be. The primary reason is that companies execute software-centric methodologies. But what makes innovation difficult is people, so companies need to execute a people-centric methodology.

A poor methodology ruins projects, jobs, and relationships, whereas a sound methodology nourishes them. Methodology matters because organizations cannot rely on heroes. Heroes come and go, but what's left is a team. Innovation is not a solo sport for heroes but a team sport, and teams need a system—a methodology—to survive changes in leadership, followership, and relationships. A good methodology is good leadership at scale.

Relationships and employee experiences improve when innovation teams overcome their addiction to meeting gridlock and email overload. Healthy teams replace that addiction with a steady rhythm of non-email documentation—that is, an asset portfolio. This asset portfolio is a vital ingredient of the Innovation Elegance methodology.

Meetings and email are easy and addictive because of their low upfront cost. Their culture works when innovation traffic is light, and the pace is slow. But when innovation traffic is heavy and the pace is hurried, relying on meetings and emails *causes* traffic jams and *slows* all the frantic moving parts. Their high marginal cost equates to laboriousness, rework, and fatigue. Weeks after a meeting occurs or an email circulates, its value depreciates. Months later, its value dissipates completely because it is *disposable*.

In contrast, proper documentation is valuable for months and years—it's *durable*. *Creating* documentation is unpopular because of its high upfront cost. *Benefiting* from documentation, however, is popular because this form of information sharing has a low marginal cost. The asset portfolio is helpful to your future team—it resembles a love letter to them. It is a ticket *out* of a traffic jam. It's not more work; it is *the* work.

Many companies organize work using an infinite number of tasks and phrases that reinvent the wheel and create a culture of Verb Sprawl. The asset portfolio shows how work can be organized with approximately sixty assets and strictly Five Verbs. The asset portfolio undermines any team's tendency to reinvent the wheel. The simplicity and low ambiguity of Five Verbs—draft, review, revise, approve, and distribute—impose a culture of discipline and empathy. The Five Verbs framework governs a kind of agreement factory, embracing the competition of ideas and the collaboration of people.

Businesses want speed, including in their innovation work. But the speed of individuals is different from that of teams. Speed in individuals is franticness. Speed in teams is synchronization, that is, synchronization of a team's activities—traffic. Synchronization of traffic improves speed, sustainability, and morale. Synchronizing an infinite number of meetings and emails is impossible and resembles a traffic jam. But synchronizing Five Verbs and five dozen assets is manageable, and succeeding at it brings out a team's best—including its speed.

Without an asset portfolio, an innovation team is trapped in a kind of debt—documentation debt—that slows innovation and hurts profits. Without a healthy methodology, an innovation team is trapped in Methodology Debt—which slows innovation and hurts profits. Executing a healthy methodology with an asset portfolio gets a team out of debt. It places you at your innovation frontier—free to explore, imagine, and pursue new paths of profit and stakeholder value.

Elegance is a healthy innovation methodology for the twenty-first century. It overcomes franticness, fatigue, and burnout and replaces them with a culture of deep discipline and empathy, that is, ruthlessness and grace.

As you probably know, many people look at the reviews on Amazon before they decide to purchase a book. If you liked this book, please leave a review with your feedback. It takes only a few minutes, and it helps future readers.

I want to know in what ways this book was helpful to you and your team. Please find me on LinkedIn at https://www.linkedin.com/company/innovation-elegance/ where you can learn more about how I apply this methodology to all sorts of ensembles.

Thank you very much,
Robert Snyder

Author Biography

Author Robert Snyder is the founder and president of Innovation Elegance, LLC. Robert's thirty-year career spans roles such as developer, project management, change management, sales enablement, and the performing arts. His career path includes corporate roles, consulting roles, startups, PMP, and Agile certifications. He's performed in numerous vocal, dance, and theater ensembles. Robert earned his BS in Electrical Engineering from the University of Illinois and his MBA in Strategy from the Kellogg School of Management at Northwestern University.

www.ingramcontent.com/pod-product-compliance
Lightning Source LLC
Chambersburg PA
CBHW031843200326
41597CB00012B/243